Sliding Mode Control for Synchronous Electric Drives

Sliding Mode Control for Synchronous Electric Drives

Sergey Ryvkin
Russian Academy of Sciences,
Trapeznikov Institute of Control Sciences, Moscow, Russia

Eduardo Palomar Lever
University of Guadalajara, CUCEI, Guadalajara, México

CRC Press
Taylor & Francis Group
Boca Raton London New York

CRC Press is an imprint of the
Taylor & Francis Group, an **informa** business

A BALKEMA BOOK

CRC Press
Taylor & Francis Group
6000 Broken Sound Parkway NW, Suite 300
Boca Raton, FL 33487-2742

First issued in paperback 2019

© 2012 by Taylor & Francis Group, LLC
CRC Press is an imprint of Taylor & Francis Group, an Informa business

Typeset by MPS Limited, a Macmillan Company, Chennai, India
No claim to original U.S. Government works

ISBN-13: 978-0-415-69038-6 (hbk)
ISBN-13: 978-0-367-38213-1 (pbk)

Library of Congress Cataloging-in-Publication Data

Sliding mode control for synchronous electric drives / Sergey Ryvkin, Eduardo Palomar Lever.
    p. cm.
  Includes bibliographical references and index.
  ISBN 978-0-415-69038-6 (hardback : alk. paper) 1. Electric motors, Synchronous—Automatic control. 2. Electric motors—Electronic control. 3. Sliding mode control. I. Ryvkin, Sergey II. Palomar Lever, Eduardo.

  TK2787.S65 2012
  621.46—dc23

                                        2011034523

Visit the Taylor & Francis Web site at
http://www.taylorandfrancis.com

and the CRC Press Web site at
http://www.crcpress.com

Contents

Preface

This book was conceived for specialists and postgraduate students who work in the field of electrical drive control. It is also ideal for the specialists in control theory application, and in particular, sliding mode using for control of electrical motors and power converters. The authors intend to leave easily available reading material besides published articles and doctoral theses.

The readers are presented with the theory of control systems with sliding mode applied to electrical motors and power converters. They can learn the methodology of control design and original algorithms of control and observation.

This book is important from a practical viewpoint, because nowadays practically all semiconductors devices are used in power electronics as power switches. Switching possesses myriad attractive inherent properties from a control viewpoint. Sliding mode control systems supplies high dynamics to systems, invariability of systems to changes of parameters and exterior loads in the combination with simplicity of design. Unlike linear control, switching sliding mode control does not replace the control system, but uses the natural properties of the control plant system effectively to ensure high control quality.

There are very few available books on using sliding mode control for electrical drives. Unlike other similar books, this book examines in detail the different features of various types of synchronous machines and converters from the viewpoint of sliding mode control design.

The manuscript presents a meticulous and detailed analysis of control issues and mechanical coordinate observation design for the various types of synchronous machines and various drive control structures. The problem of drive parameters identification is discussed as well.

The potential of the sliding mode control and observation is demonstrated in numerical and experimental results for real control plants.

The authors

Acknowledgements and Dedications

To my son Denis

It is a pleasure for me to acknowledge the contributions of my former University lecturer Prof. V. Polkovnikov and my former Ph.D. supervisor Prof. V. Utkin, who made me a scientist.

I wish to express my sincere gratitude to my colleagues at the Trapeznikov Institute of Control Sciences of Russian Academy of Sciences and particularly to Academician Prof. S. Vassilyev, Prof. A. Shubladze, Prof. V. Lotozky and Dr. D. Izosimov. Their support was invaluable on the different writing up phases.

Thanks to Prof. Y. Rozanov, IEEE Fellow, for his important advice and remarks when preparing the final manuscript.

I would also like to express my sincerest thanks to Mrs. M. Platova. Without her assistance, this book could not have been started and completed.

Sergey Ryvkin

To my beloved parents Javier and Bertha

I would like to express a word of thanks to Dr. P.K. Sinha, for his invaluable help during my postgraduate studies, and to one of my former university lecturers in Mexico, Prof. Felipe Gálvez, who made me a control engineer.

I also would like to express my deepest thankfulness to my adored parents, Javier (†) and Bertha, an example of moral principles and behavior, hard work, dedication and love, and to whom I owe everything.

Eduardo Palomar Lever

We would like this book to be free of errors, even if we know that this is impossible in practical terms. So, we would appreciate if you could send any error reports to us to the following email addresses:
sergey_ryvkin@mail.ru, palomare@mail.ru

About the authors

Sergey Ryvkin first graduated with high honors as an engineer from the Moscow Institute for Aviation Engineering (Technical University), after which he gained his PhD degree from the Institute of Control Sciences (USSR Academy of Science) in Moscow and was awarded a DSc from the Supreme Certifying Commission of Russian Ministry of Education and Science in Moscow. He is currently a professor at the Russian State University for Humanities and a leading researcher at the Laboratory of Adaptive Control Systems for Dynamic objects at the Trapeznikov Institute of Control Sciences from the Russian Academy of Sciences. His lines of research are the application of the sliding mode techniques to control of electrical drives and power systems and to their parameter observation. Prof. Ryvkin holds several patents and published one monograph, five textbooks and more than 100 papers in international journals and proceedings. He is a member of the Russian Academy of Electrotechnical Sciences and a senior member of the IEEE.

Eduardo Palomar Lever got his BSc degree in electromechanical engineering from the National Autonomous University of Mexico, after which he obtained an MSc degree in control engineering and computing sciences from the University of Warwick, UK, and a PhD on sliding regimes to control servomechanisms from the University of Sussex, UK. He is a full time research professor at the University of Guadalajara. His lines of research focus on nonlinear control systems and servomechanisms control using digital sliding modes. Prof. Palomar-Lever's teaching specialties are advanced-level control engineering, automation, advanced mathematics, computing languages, software development, and statistics. He earlier published a book on ferroelectric materials as well as several papers on sliding motion control and biomedicine in international journals and conference proceedings.

Introduction

Nowadays, while automating technological processes, there is a tendency to use the general principles of control system design in difficult cases, in which the maximum order considered is outlined and the properties of the plants are used. In the frame of such approach, the specific properties of the plants, due to their physical nature, are simply not considered, but effectively used for goal achievement. This way, new principles and laws of control, along with maintenance, make a high quality of system control possible, and provide high technical and economic indicators. However, in the nonlinear world, there are no standard ways or universal methods, which are characteristic of the linear control theory. Each nonlinearity is different and has several design methods.

Among nonlinear control plants, undoubtedly, there is one kind that occupies a leading place from the viewpoint of electrical power consumption (Suto & Nagy, 2006). It requires more than 60% of all the consumed electrical energy produced in industrially developed countries (Leonhard, 2001), (Benda, 1996), (Bose, 2002), (Mohan, 2003) (Szentirmai, 2000), (Blaabjerg et al, 2010), (Ohashi, 2010). About one third of the primary energy, often not renewable, goes today into electrical power. Current consumption is increasing all the time. Therefore, a very real problem is the rational use of electric power, i.e. goal achievement with minimum energy expenses. One real example is a drive control problem, i.e. working out drive controls, which would provide the desired performance with minimum loss of electric power (Harasnima & Hashimoto, 1986), (Izosimov & Ryvkin, 1986), (Kuerker, 2000).

Today, the best perspectives from the viewpoint of efficiency and compactness of a drive control design have an alternating current motor fed by semiconductor power converters. The characteristics feature of such drives is relay nonlinearities. Those modern power devices work mainly in "a switching mode" for the purpose of achieving small power losses (Benda, 1996), (Bose, 2003), (Mohan, 2003). Such dynamic systems are essentially nonlinear, and they are described by their differential equations with discontinuous controls. This is why such systems are called "relay systems" or "systems with discontinuous control".

The history of relay systems is much more ancient than the history of semiconductor devices, and it begins with the relay feedback coupling used by C. Schofield in 1836. Despite propensity of relay systems to self-oscillations in the area of zero control error, simplicity of their implementation combined with high dynamic properties, and their property of self-adapting to parameter changes and different loads have provided such systems with a wide popularity, so the creation and development of the theory of relay systems came as a consequence.

The first stage of development of relay system theory is connected first of all with the names of H. Hazen (Hazen, 1934), A. Andronov (Andronov, 1959), Y. Tsypkin (Tsypkin, 1974) and I. Flugge-Lotz (Flugge-Lotz, 1953). Further from the relay system theory, the following directions were declared as independent:

- The theory of nonlinear systems with various kinds of modulation (Kuerker, 2000), (Holz, 1994) and
- The variable structure system theory (Emelyanov, 1970), (Edwards & Spurgeon, 1998).

The founder of the last one is Prof. Stanislav Emelianov. He headed a variable structure system academic school, whose scientists brought an essential contribution to this theory. The basic idea of this theory is the use of sliding modes in control design. Sliding mode is a special kind of movement arising under certain conditions in switching systems and inherent only to them. The specified mode provides in a dynamic system a high quality of control, invariance to external unmeasured disturbances, and small sensitivity to changes of dynamic properties of the plant.

The variable structure system theory was further developed and generalized with the theory of systems with discontinuous control (Utkin, 1978), (Utkin, 1992), (Edwards & Spurgeon, 1998), (Utkin et al, 1999), the theory of binary systems (Emelyanov & Korovin, 2000) and the theory of higher order sliding modes (Fridman & Levant, 2002). The theory of systems with discontinuous control is based on the use of a multidimensional sliding mode in the state space for the solution of control problems. The theory of binary systems is based on binary principles, i.e. the dual nature of signals in nonlinear dynamic systems that allows assigning operation design using stabilizing feedback to an auxiliary nonlinear system. The theory of higher order sliding modes generalizes the basic sliding mode acting on higher order time derivatives of the system, which totally removes the chattering effect and provides high accuracy in its implementation.

Possibility and perspectives of the use of sliding modes for alternating current drive control design was formulated for the first time in (Sabanovich & Izosimov, 1981) though relay controllers found wide application in drive control before (Il'insky, 2003). It is remarkable to note that, independently of the experts in the control field, the experts in the drive field also have addressed the use of relay control on the basis of sliding modes (Brodovsky & Ivanov, 1974). These controls were used in the phase current loops. Their use has been encouraged by the advances of new semiconductor technologies and transition to power semiconductor voltage or current converters, whose power elements work in switching (relay) modes.

Rapid development of power semiconductor technologies has led to the development of new types of high-frequency power devices, such as MOSFET and IGBT, which opened plenty of opportunities to create and perfect semiconductor power converters for drives (Benda, 1996), (Bose, 2002), (Mohan et al, 2003). All of the multidimensional relay control more actively used in the last decade worked mainly in phase current loops of the drive. The increasing number of publications testifies it. Such control in various publications has different names: "relay control" (Il'insky, 2003), "discontinuous control" (Borcov, 1986), "frequency-current control" (Brodovky & Ivanov, 1974), "sliding mode control" (Utkin et al, 1999), (Il'insky,

2003), (Cernat et al, 2000), (Vittek & Dodds, 2003), "bang-bang control" (Isidori, 1999), "hysteresis current control" (Holz, 1994), (Holz & Beyer, 1994), "current forced control" (Nagy, 1994), (Suetz et al, 1996), "direct torque control" (Leonard, 2001), (Buja & Kazmierkowski, 2004), (Lascu et al, 2004), etc. Such variety of names for one kind of control testifies that experts in the drive field do not have a common viewpoint as to where and how to place the considered approach into the available control methods for the drives. The majority of the available publications are devoted, as a rule, to disclosing private questions of research and implementation of drives with such control.

The term "sliding mode control" is preferred, in our opinion, because it provides the best and fullest explanation of the processes involving the use of this kind of control. There is a whole theory of nonlinear systems with discontinuous controls behind this term. This theory widely explains not only the known high quality results obtained by the use of this kind of control, but also those problems and complexities that arise at its implementation.

The difficulty of application of the sliding mode approach to the drive control design is that the methodology of the considered approach is highly theoretical. It uses mathematical models of the control plants that look like differential equations with discontinuous control (Filippov, 1998), (Aizerman & Pyatnitskii, 1974). Besides that, direct use of the theory for drive control design is impossible without additional research on questions concerning the organization of sliding modes, taking into account specific features of the drive elements: electrical machines, semiconductor power converters, sensors, etc. Though, as specified above, it is quite natural from the viewpoint of the physical processes proceeding in the drive using the theory of systems with discontinuous controls. Motor controls are voltages in the stator windings of the electric machine. They have a discontinuous character, owing to switching kind of work of the semiconductor elements of the voltage converter. The discontinuous character of the control, in this case, which is the defining sign of the theory of nonlinear systems with discontinuous control, is not imposed to the system as an outside property, but it is naturally defined by its physical nature.

The multidimensional relay characteristic of the power converter, which is defined by the drive control design is not unique. It is necessary to consider the nonlinearity of alternating current machines. In each of two alternating current electric machines, i.e. asynchronous and synchronous, the process of transformation of electric power into mechanical one has essential differences. It is caused by the basic distinction in a source of a magnetic flux in the air gap, necessary for the creation of electromagnetic torque. The stator current generates this flux in the asynchronous motor using the electromagnetic induction. However, in the synchronous machine, it is generated independently by a flux source located in the rotor. Considering that the synchronous machine combines very attractive properties as small rotor losses and good dynamic and accuracy characteristics, and in view of the fact that nonlinear characteristics of the former make an essential impact on drive control design, basic attention is given to problems of synchronous drive control design in the present monograph. The semiconductor power converter and the synchronous motor enter into the drive structure.

Thus, a three-phase synchronous drive represents a nonlinear dynamic system with the linear occurrence of the control $u(t)$ with discontinuous character, caused by the switching operation mode of the elements of the power converter.

The following prominent features of this class of nonlinear dynamic systems with discontinuous control, in comparison with the widely researched traditional one (Utkin, 1978), (Utkin, 1992), are the following:

- The amount of discontinuous control surpasses the dimension of control space (three-phase feed voltage of the electric machine at a two-dimensional voltage vector);
- Basic vectors (orts) of the discontinuous controls used for the solution of a control problem are fixed;
- Coefficients before the discontinuous controls are periodic (in the case of salient-pole (interior) synchronous motors).

Development of the theory of nonlinear systems with discontinuous controls to such class of nonlinear systems has allowed developing design methods of nonlinear control on sliding modes for the given concrete class of systems taking into account its features, i.e. to physically use as much as possible potential possibilities for control problem solving. With reference to three-phase drive, it means high quality of control processes, invariance to external perturbations, small sensitivity to changes of dynamic properties of plant, in a combination to profitability of power transmission and simplicity of reception of the rotating magnetic field, inherent in three-phase circuits.

Realization of the high-quality control based on use of a multidimensional sliding mode is impossible without due information support, which consists in reception of the necessary information on state vector components of the plant. Direct measurement of all components of state vector needed for control design is impractical due to essential complications, cost issues and reduction of the plant operational reliability. A perspective way of the solving of information support problem is working out of estimation algorithms for all the unmeasured components of the state vector based on the observed components (Consoli, 2000), (Dote, 1988), (Krstic et al, 1995), (Luenberger, 1966).

From the viewpoint of sliding mode the problem of the estimating algorithms design has two parts. One part consists in receiving the state vector component estimates needed for the sliding mode control design. Another part deals with using sliding mode techniques for receiving these estimates. In the latter case, methods of nonlinear observation are based on the construction of a dynamic, imitating model of a nonlinear plant with discontinuous modeling control and the use of the attractive property of sliding movement, which is the possibility of allocating an average continuous value of a discontinuous control as a control information signal.

The present monograph contains sliding mode methods of control and observation design for nonlinear systems with a periodic matrix before redundant discontinuous control, developed with uniform positions. Such approach allows making the most of their structural features for achievement of the control goal. The offered approach with reference to drives has allowed to develop design methods of the high-quality information provided controls, both in continuous, and in discrete time. The synthesized controls fully use the physical nature of drive elements for a control problem. They are characterized by a high quality of control, invariance to external perturbations, small sensitivity to changes of feed voltage and dynamic properties of the synchronous motor. They provide a high degree of usability of energy in a combination to profitability of transmission of energy and simplicity of reception of the rotating magnetic field, inherent in the three-phase circuits.

The control tasks considered in this book found their reflection in the monograph in content and structure, which consists of an introduction and eight chapters.

In chapter 1, the basic elements of the automated synchronous drive are classified from a position of the automatic control theory as power converters and synchronous motors. The mathematical descriptions of the power converters and the synchronous motors are exposed. Control problems are formulated and formalized.

In section 1.1 classifications of the synchronous motors by a principle of a magnetic flux creation and of the power converters by a principle of transformation of input voltage into a three-phase alternating voltage of the set frequency and amplitude are presented. The mathematical models used for the control problem solving are shown.

In section 1.2, the basic requirements to drives are formulated and various drives structures and features of transformation of initial requirements depending on drive structures are considered.

In chapter 2, theoretical backgrounds of multidimensional sliding movement design in nonlinear dynamic systems with a periodic matrix before redundant discontinuous control are stated. From uniform positions of the sliding mode theory control design problems are solved for a considered class of nonlinear systems. The sufficient existence condition of a sliding mode is formulated and proved. This condition is the base for a two step control design procedure.

In section 2.1, features of a considered class of the nonlinear dynamic systems are analyzed. It is explained why the direct use of classical results of the theory of systems with discontinuous controls for the concerned dynamics systems is not possible. Results of this classical sliding mode theory used in this work are produced.

In section 2.2, the sufficient existence conditions of a sliding mode in the systems under research are formulated and proved. The special case having a key importance for the control design for the three-phase drives is considered.

A two-step procedure of a sliding mode control design for the investigated systems is presented to section 2.3.

Chapter 3 is devoted to problems of information support of existence of multidimensional sliding movement for nonlinear dynamic systems at a limited number of the measured component of a state vector. Use of asymptotical observers, as information basis of the sliding movement design, providing invariance of sliding movement to discrepancies of model and measurements in high frequency areas is proved. The solution of a problem of nonlinear estimation state vector components by using sliding mode in a special nonlinear dynamic model with discontinuous controls is offered.

In section 3.1, information aspects of the organization of a sliding mode are discussed. It is shown that from the information viewpoint it is possible to allocate two kinds of problems associated with sliding motion. First, the reception of the information on state variables, necessary for realization of sliding motion in a control loop. Second, the use of sliding motion for reception of the necessary information on state variables.

In section 3.2, it is shown that the use of asymptotical observers is a methodological basis of a sliding mode design. Their use allows removing a problem of sensitivity of sliding movement in relation to high-frequency non-ideal behavior.

Section 3.3 is devoted to working out problems of design methods of nonlinear observers based on sliding modes for nonlinear systems with a linear occurrence of the estimated state vector components. Sufficient conditions of nonlinear estimation of the state vector components are formulated and proved. The observation algorithms allow essentially reducing the demanded number of calculations.

Chapter 4 is devoted to the problem of working out control design methods and algorithms of automated synchronous drive with use of sliding modes for cases of one-loop and cascade control of drive and various types of synchronous motors and power converters. A two-step decomposition design procedure of control allowed separately considering by drive control design features of a synchronous motor and a supplied power converter is presented. Problems of maintenance of drive invariance to changes of plant parameters, external perturbations and changes of feed voltage are discussed. Also, the problem of creating the reference value of a stator current component i_{dz} based on the drive technical and economic requirements is discussed.

In section 4.1, a decomposed two-step design method of a one-loop control is described. On the first step of the control design only the features of the synchronous motor in the rotating coordinates frame (d, q) is carried out. On the second step, the features of the used power converter are considered and the phase voltage controls are synthesized.

Section 4.2 presents the use of the above mentioned approach for design of the cascade control when the voltage source inverter works as a current source inverter.

Section 4.3 is devoted the solving of the optimization problems of drive power characteristics by using the formation of the reference value of the stator current component i_{dz} in the closed loop. The reference value is formatted so that either it achieves maximization of efficiency or it does not use reactive power.

Chapter 5 is devoted to working out design methods of discontinuous control in systems with a multidimensional real sliding mode. The problems arising at use of sliding modes for control in this case are analyzed. The controls providing a regularity commutating discontinuous component of control at the expense of a special choice of switching surfaces are synthesized. As an example, a current control problem of the synchronous motor fed by the voltage source inverter is solved. The offered approach allows providing high dynamic and accuracy of the system in a combination with minimum switching losses and performance of electromagnetic compatibility requirements.

In section 5.1 the features of a real sliding mode caused by final frequency of switching of power devices and non-ideal of their relay characteristics are analyzed. Methods to regularize switching frequencies in a real sliding mode are planned.

In section 5.2 problems of optimization of feedforward (program) control by the voltage source inverter on the basis of PWM, using switching losses minimization as a criterion, are solved. The optimal PWM for the voltage source inverter, based on allocation of zones of optimality, is synthesized with use of two step comparative analyzes of possible PWMs by criterion of switching losses minimization. Implementation problems are discussed.

In section 5.3 the method of control design providing realization of feedback PWM is developed, whose properties are equivalent to those of a feed forward PWM, i.e. the resulting real sliding mode is optimal on switching losses.

In section 5.4 the method of the control design providing regularization of the discontinuous switching component of a control vector is developed.

Chapter 6 is devoted the estimation problems of drive output mechanical coordinates under the current information on drive electric variables. On the basis of results of chapter 3, design methods of nonlinear state observers based on sliding modes are offered. Estimation algorithms of mechanical coordinates for exterior synchronous motor and the synchronous reluctance one are designed. The features of various observation algorithms are discussed. The block diagrams of nonlinear state observers are resulted.

In section 6.1 the general statement of an observation problem of drive mechanical variables is discussed. The features of a synchronous motor, as nonlinear plant are underlined. It is emphasized that owing to specificity of these nonlinear plants, the design problem of an observation algorithm of mechanical coordinates must be solved separately for each type of synchronous motor. Variables accessible to measurement are allocated.

Section 6.2 presents the sliding mode observer for the exterior synchronous motor with the permanent magnet excitation. The main idea is used as an observer a special dynamics system, in which the sliding mode is willfully organized. The average values of the discontinuous controls give the information about the mechanical coordinates. Observation algorithms of variables in the stationer and rotating coordinates frames are produced.

In section 6.3 the above mentioned approach is used for the synchronous reluctance motor. Observation algorithms are produced.

Chapter 7 is devoted to design problems of a digital control for the synchronous drive. Features of organization of digital control and a realization of a sliding mode in such systems are considered. The design methods of digital control and observation for synchronous drives, guaranteeing certainly step or asymptotical character of the processes are developed. Features of use of various control and observation are discussed. For the purpose of simplification of control design procedure, the approach based on use of the reference rate limiter is offered. It provides an exception of variable constrains in the course of system functioning. The design method of such limiters is developed. Conditions of system parameter identification that connect a depth of memory, a frequency of quantization and a quantity of identified parameters are formulated. The problem of identification of the moment of inertia of the synchronous motor is solved as an example. The condition of its identification is formulated. Problems of the digital control of the drive with flexible joints are considered. The condition of oscillatory free movements is formulated. The control for such drive is synthesized.

Features of digital control and organization of sliding movement are considered in section 7.1.

Digital control design methods, based on difference equations, are developed in section 7.2.

The developed design methods of digital variable estimation under the retrospective and current information are presented in section 7.3. The novel algorithms of estimation and a filtration of drive output mechanical variables are developed.

The developed methods of design of digital algorithms of identification of system parameters are presented in section 7.4. The algorithm of identification of the moment of inertia of a motor is developed.

The developed design methods of digital algorithms of the reference rate limitation are presented in section 7.5. They provide an exception of mechanical coordinates constrains by the control design.

The developed design method of digital control of the drive with elastic mechanical joints and the control synthesized with its use is presented in section 7.6.

Chapter 8 presents the results of the using the above supposed controls and observation algorithms for the control of the different technological processes. Features of use and technical realization of such control systems are discussed.

The solving design problem of digital control of the high speed synchronous drive with a vector digital control without the mechanical coordinate's sensors on a motor shaft is presented in section 8.1. The problem was a part of the federal target program called "National technological base". The subject was "Prototyping a small-sized high speed drive for the oil drowned pumps for oil extracting with the inverter and microprocessor control of power to 200 kW" and the code of the subject was "Nupor". (The directing agency is the federal state unitary enterprise "Andronicus Josephian Research and Production Enterprise All Russia Scientific Electro-Mechanics Research Institute"). The features of a solved problem imposing special requirements on control and approaches used are described. Modeling results have confirmed efficiency of our approach to digital control design.

Section 8.2 presents the digital control design for a drive with elastic mechanical joints, developed for the state unitary enterprise "Instrument Design Bureau" (State Unitary Enterprise KBP). The name of developmental work was "Prompting and stabilization drives of special plant". The code was "Punzir", under the subject "Working out an alternating current drive with a vector and adaptive-modal microprocessor control". The features of the solved problem and used approaches are described. The modeling results have confirmed efficiency of our approach to designing digital control.

REFERENCES

Aizerman, M.A. and Pyatnitskii, E.S. *"Fundamentals of a theory of discontinuous systems."* 2 parts. Automation and Remote Control, 1974, vol. 35, part 1, no. 7-1, pp. 1066–1079, part 2, no. 8-1, pp. 1242–1261.

Andronov, A., Witt, A. and Hikin, S. Oscillation theory. Moscow: Physmathgiz, 1959. 916 p. (in Russian).

Benda V. *"Reliability of power semiconductor devices – Problems and trends"*. Proc. 7th International Power Electronics & Motion Control Conference, PEMC'96, Budapest, Hungary, 1996, vol. 1, pp. 30–35.

Blaabjerg, F., Iov, F., Kerekes, T. and Teodorescu, R. *"Trends in power electronics and control of renewable energy systems"*. Proc. 14th International Power Electronics and Motion Control Conference, EPE-PEMC 2010, Ohrid, Republic of Macedonia, 2010, pp. K-1–K-19.

Borcov, Ju. and Junger, I. *"Automatic systems with discontinuous controls"*. Leningrad: Energoizdat, 1986. 168 p. (in Russian).

Bose, B.K. *"Modern power electronics and AC drives"*. New Jersey: Prentice Hall, 2002. 711 p.

Brodovsky, V. and Ivanov, E. *"Frequency-current control drives"*. Moscow: Energy, 1974. 168 p. (in Russian).

Buja, G.S. and Kazmierkowski, M.P. "*Direct torque control of PWM inverter-fed AC motor – a survey*", IEEE Transactions on Industrial Electronics, 2004, vol. 51, pp. 744–757.

Cernat, M., Comnac, V., Cotorogea, M., Korondi, P., Ryvkin, S. and Cernat, R.M. "*Sliding mode control of interior permanent magnet synchronous motors*". Proc. the 7th IEEE Power Electronics Congress, CIEP 2000. Acapulco, Mexico, 2000, pp. 48–53.

Consoli, A. "*Advanced control techniques*". Modern Electrical Drives. Dordrecht, Boston, London: Kluwer Academic Publishers, 2000, pp. 523–582.

Dote, Y. "*Application of modern control techniques to motor control.*" Proc. of the IEEE, 1988, vol. 76, no. 4, pp. 438–445.

Edwards, C. and Spurgeon, S.R. "*Sliding mode control: theory and applications*". London: Taylor & Francis, 1998. 237 p.

Emelyanov, S. and Korovin, S. "*Control of complex and uncertain systems*". London: Springer-Verlag Ltd, 2000. 332 p.

Emelyanov, S., Utkin, V., Taran, V., Kostyleva, N., Shubladze, A., Eserov, V. and Dubrovski, E. "*Theory of variable structure control systems*". Moscow: Nauka, 1970. 592 p. (in Russian).

Filippov, F. "*Differential equations with discontinuous righthand sides*". Dordrecht: Kluwer Academic Publishers, 1988. 304 p.

Flugge-Lotz, I. "*Discontinuous automatic control*". Princeton, New Jersey: Princeton Univ. Press, 1953. 150 p.

Fridman, L. and Levant, A. "*Higher order sliding modes.*" in: Sliding Mode Control in Engineering, J.P. Barbot, W. Perruguetti (Eds.), New York: Marcel Dekker, 2002, pp. 53–101.

Harasnima, F. and Hashimoto, H. "*Variable structure strategy in motion control*". Proc. Conference on Applied Motion Control, Minneapolis, Minnesota, USA, 1986, pp. 191–198.

Hazen, H. "*Theory of servomechanisms*". Journal of the Franklin Institute, 1934, vol. 218, no. 3, pp. 279–331.

Holtz, J. and Beyer, B. "*Optimal pulsewidth modulation for AC servos and low-cost industrial drive*". IEEE Transactions on Industrial Applications, 1994, vol. 30, no. 4, pp. 1039–1047.

Holz, J. "*Pulsewidth modulation for electronic power conversion*". Proc. of the IEEE, 1994, vol. 82, no. 8, pp. 1194–1213.

Il'insky, N. "*Drive backgrounds*". Moscow: Publishing house of Moscow power engineering institute, 2003. 221 p. (in Russian).

Isidori, A. "*Nonlinear control systems II*". Berlin: Springer-Verlag, 1999. 293 p.

Izosimov, D. and Ryvkin, S. "*Improvement in the quality of energy consumption using semiconductor converters with pulse-width modulation*". Electrical Technology, 1996, no. 2, pp. 33–46.

Krstic, M., Kanellakopoulos, I. and Kokotovic, P. "*Nonlinear and Adaptive Control Design*". New York: Wiley, 1995. 563 p.

Kuerker, O. "*Modulation techniques*". Modern Electrical Drives. Dordrecht, Boston, London: Kluwer Academic Publishers, 2000, pp. 289–310.

Kwakernaak, H. and Sivan, R. "*Linear optimal control systems*". New York: John Wiley & Son Inc., 1972. 608 p.

Lascu, C., Boldea, I. and Blaabjerg, F. "*Variable-structure direct torque control – a class of fast and robust controllers for induction machine drive*". IEEE Transactions on Industrial Electronics, 2004, vol. 51, pp. 785–792.

Leonhard, W. "*Control of electrical drives*". Berlin: Springer-Verlag, 2001. 460 p.

Luenberger, D.C. "*Observers for multivariable systems*". IEEE Transactions on Automatic Control, 1966, vol. 11, no. 1, pp. 190–197.

Mohan N., Underland, T.M. and Robbins, W.P. "*Power electronics: converters, applications and design*". 3rd ed. New York: John Wiley & Son Inc., 2003. 824 p.

Nagy, I. "*Improved current controller for PWM inverter drives with the background of chaotic dynamics*". Proc. the 20th International Conference on Industrial Electronics Control and Instrumentation, IECOM'94, Bologna, Italy, 1994, pp. 561–566.

Ohashi, H. "*Role of green electronics in low carbonated society toward 2030*". Proc. 14th International Power Electronics and Motion Control Conference, EPE-PEMC 2010, Ohrid, Republic of Macedonia, 2010, pp. K-20–K-25.

Pfaff, G. and Wick, A. "*Direkte Stromregelung bei Drehstromantrieben mit Pulswechselrichter*". Regelungstechnische Praxis, 1983, Bd. 4, N. 11, S. 472–477.

Ryvkin, S. "*Sliding mode technique for AC drive*". Proc. the 10th International Power Electronics & Motion Control Conference, EPE – PEMC 2002 Dubrovnik & Cavtat, Croatia, 2002, p. 444 & CD-ROM.

Sabanovic, A., Jezernik, K. and Sabanovic, N. "*Sliding mode applications in power electronics and electrical drives*". Variable Structure Systems: Towards the 21 Century, Berlin: Springer-Verlag, 2002, pp. 223–252.

Sabanovich, A. and Izosimov, D. "*Application of sliding modes to induction motor control*". IEEE Transactions on Industrial Applications, 1981, vol. 17, no. 1, pp. 41–49.

Suetz, Z., Nagy, I., Backhauz, L. and Zaban, K. "*Controlling chaos in current forced induction motor*". Proc. the 7th International Power Electronics & Motion Control Conference, PEMC'96, Budapest, Hungary, 1996, vol. 3, pp. 282–286.

Suto, Z. and Nagy, I. "*Nonlinearity in control electrical drive: review*". Proc. IEEE International Symposium on Industrial Electronics ISIE 2006, Montreal, July 2006, pp. 2069–2076.

Szentirmai, L. "*Considerations on industrial drives. Modern Electrical Drives*". Dordrecht, Boston, London: Kluwer Academic Publishers, 2000, pp. 687–722.

Tsypkin, Y. "*Theory of relay control system*". Moscow: Nauka, 1974. 575 p. (in Russian).

Utkin, V. "*Sliding modes and their application in variable structure systems*". Moscow: Mir Publ., 1978. 368 p.

Utkin, V. "*Sliding modes in control and optimization*". Berlin: Springer-Verlag, 1992. 286 p.

Utkin, V., Shi, J. and Gulder, J., "*Sliding modes in electromechanical systems*". London: Taylor & Francis, 1999. 344 p.

Vittek, J. and Dodds, S.J. "*Forced dynamics control of electric drives*". EDIS – Publishing Center of Zilina University, Slovakia, 2003. 356 p.

Zinober, A.S. "*Variable structure and Lyapunov control*". Berlin: Springer-Verlag, 1994. 420 p.

Chapter 1

Problem statement

1.1 MATHEMATICAL MODELS OF THE DRIVE ELEMENTS

1.1.1 Synchronous motors

Any electrical motor is a power transformer that converts electrical energy in mechanical one. When it is in action it works under the principle of interaction of currents and magnetic fields, which leads to a general analytical description of the processes occurring in any motor, including the present work. Because of the complexity of the electromagnetic and electromechanical processes happening in the real electric machine, its full mathematical description is impractical, and from the control viewpoint it is ineffective due to the "Dimension Curse", well known in classical control. It is sufficient to use the classical mathematical models that consider the basic physical features of the processes happening inside the motor for control design (Leonhard, 2001), (Boldea & Nasar, 2005). These mathematical descriptions are based on the following standard assumptions:

- The magnetomotive forces created by phase currents have sinusoidal distribution along an air gap, i.e. influence of the higher frequency harmonics of a magnetic field is not considered;
- Symmetry of the electrical machine;
- Influence of grooves is not considered, but the machine can be salient-pole;
- Absence of saturation and losses in steel;
- The energy of any electrostatic field is not considered;
- Processes are concentrated.

In the frame of these assumptions the mathematical description of any electric motor includes following three groups of equations:

- Equations of electric balance in its windings;
- Equation of the electromagnetic torque developed by the electric motor;
- Equation of mechanical movement (Newton's second law for a rotary motion).

The third equation, i.e. the equation of mechanical movement, is general for all electric motors and looks like this:

$$J\frac{d\Omega}{dt} = M_{el} - M \tag{1.1}$$

where J is the moment of inertia of all rotating parts, reduced to an motor rotor; Ω is the angular speed of the rotor; t is the current time; M_{el} is the electromagnetic torque developed by the electric motor; $M = \sum M_l(F_i)$ is the sum of the torques of the external forces enclosed to a rotor.

The equations of electric balance based on Kirchhoff's second law, and the equations of electromagnetic torque are defined by a kind of electric and magnetic circuits of the electric engine and the physical processes happening inside it. Therefore, they should be considered independently for each type of synchronous motors.

From the control design viewpoint, all varieties of synchronous motors (Hanitsch & Parspour, 2000), (Pahman & Zhou, 2000) could be broken into several types, depending on the following two criteria:

–   The rotor form, i.e. the form of an air gap in the electric engine.
–   The source of the magnetic flux causing the torque.

A classification of the types of synchronous motors based on such approach is shown in figure 1.1.

The first criterion in the given classification is the form of the air gap. There is a uniform air gap in the case of a *nonsalient-pole* rotor. Otherwise there is a non-uniform gap between the rotor and the stator in a salient pole motor.

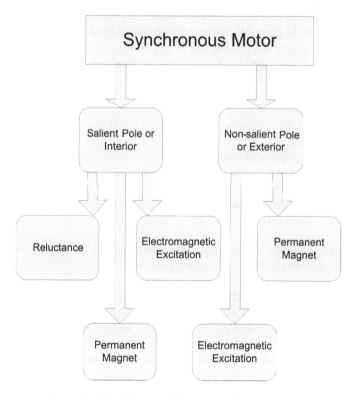

Figure 1.1 Classification of electric synchronous motors

The magnetic flux in a synchronous motor can be created by permanent magnets or a winding located on the rotor. In the latter case, sliding contacts are used to supply electric power to the rotor. The family of salient-pole synchronous motors concerns also the synchronous reluctance machine, in which there is no independent source of magnetic flux. Using the non-uniformity of the air gap forms the magnetic flux needed for torque creation.

The mathematical description of the electromechanical transformation of power in the synchronous motor using the actual phase currents and voltages as independent variables gives a direct representation of the physical processes happening in the motor. However this mathematic model is complicated enough for analysis of the dynamics processes. In the case of salient-pole machines this complexity is aggravated by fact that the synchronous motor description is a system of nonlinear differential equations with periodic coefficients.

Park's equations are used in the theory of electrical machines for research on the synchronous machine (Leonhard, 2001), (Boldea & Nasar, 2005), (Park, 1929). These equations are obtained by a linear Park's transformation of the above mentioned physical equations of the synchronous motor. The result is a set of differential equations with fictitious variables in the rotating coordinates frame. (Adequacy of the mathematical description is ensured using the condition of power invariance). This transformation provides transition from the initial fixed coordinates system connected with motor phases A, B, C to the rotating coordinates system $(0, d, q)$ connected to a rotor. The datum line d is connected to the rotor flux, and the datum line q is orthogonal to it. The datum line 0 is guided on an axis of rotation of the synchronous engine (figure 1.2).

The successful choice of transformation to axes $(0, d, q)$ plays a fundamental role in the theory of synchronous machines and the theory of automatic drives, and clearly proves the following statements. First, thanks to the rotating coordinates system connected to a rotor, the differential equations describing the synchronous motor dynamics have constant coefficients. Second, the orientation of a datum line 0 on the axis of rotation makes the differential equation independent as regards a zero sequence current. It must be emphasized that the zero sequence current does not participate in the magnetic field creation in an air gap, and, therefore, it does not have influence on the electromechanical processes in the motor, but only creates an additional loading of windings and semiconductor devices. The traditional

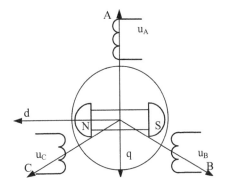

Figure 1.2 The synchronous motor

connection of motor phase windings "star" or "delta" automatically provides equality to zero of a zero sequence current, i.e. it eliminates additional thermal losses. From the viewpoint of the mathematical description of the synchronous motor it leads to its simplification. Its order is reduced after eliminating the differential equation of the zero sequence current. In this case the considered three-phase synchronous motor is replaced by a two-phase "idealized" one (the generalized electric machine), representing the collector machine, from the viewpoint of electromagnetic processes.

The transformation from the three-phase coordinates system to the rotating two-phase one (d, q) is implemented by means of a Park's transformation matrix:

$$C = \sqrt{\frac{2}{3}} \begin{pmatrix} \cos \gamma_A & \cos \gamma_B & \cos \gamma_C \\ -\sin \gamma_A & -\sin \gamma_B & -\sin \gamma_C \end{pmatrix} \tag{1.2}$$

where γ_j is an electric angle between a rotor axis d and a stator phase axis j ($j = A, B, C$) and $\sqrt{2/3}$ is a power invariance factor.

The electric variables in a rotating coordinate system i_d, i_q, u_d, u_q are also connected to phase variables:

$$I = CI', \quad U = CU' \tag{1.3}$$

where $I^T = (i_d, i_q)$, $I'^T = (i_A, i_B, i_C)$, $U^T = (u_d, u_q)$, $U'^T = (u_A, u_B, u_C)$.

The equations of electric balance and the electromagnetic moment for various types of synchronous engines are shown below:

1.1.1.1 Salient-pole synchronous motor with an excitation winding

$$\begin{aligned}
\frac{di_d}{dt} &= \frac{1}{L_1^2}(-L_f r i_d + L_{df} r_f i_f + L_f L_q \omega i_q) + \frac{L_f}{L_1^2} u_d - \frac{L_{df}}{L_1^2} u_f, \\
\frac{di_q}{dt} &= \frac{1}{L_q}(-r i_q - L_d \omega i_d - L_{df} \omega i_f) + \frac{1}{L_q} u_q, \\
\frac{di_f}{dt} &= \frac{1}{L_1^2}(-L_d r_f i_f + L_{df} r i_d - L_{df} L_q \omega i_q) + \frac{L_d}{L_1^2} u_f - \frac{L_{df}}{L_1^2} u_d, \\
M_{el} &= \frac{3}{2} p[L_{df} i_f + (L_d - L_q) i_d] i_q,
\end{aligned} \tag{1.4}$$

where i_d, i_q are the currents in the stator windings; i_f is the excitation current; u_d, u_q are the voltages in the stator windings; u_f is the voltage of the excitation winding; r is the active resistance of a stator winding; r_f is the active resistance of the excitation winding; L_d, L_q are the inductance of stator windings; L_f is the inductance of the excitation winding; L_{df} is the mutual inductance excitation windings and stator windings d; $L_1 = \sqrt{L_f L_d - L_{df}^2}$; p is the number of pair poles; ω is the electric angular speed.

1.1.1.2 Permanent magnet salient-pole synchronous motor

$$\frac{di_d}{dt} = \frac{1}{L_d}(-ri_d + L_q\omega i_q) + \frac{1}{L_d}u_d,$$

$$\frac{di_q}{dt} = \frac{1}{L_q}(-ri_q - L_d\omega i_d - \Psi_f\omega) + \frac{1}{L_q}u_q, \qquad (1.5)$$

$$M_{el} = \frac{3}{2}p[\Psi_f + (L_d - L_q)i_d]i_q$$

where Ψ_f is a excitation flux.

1.1.1.3 Synchronous reluctance motor

$$\frac{di_d}{dt} = \frac{1}{L_d}(-ri_d + L_q\omega i_q) + \frac{1}{L_d}u_d,$$

$$\frac{di_q}{dt} = \frac{1}{L_q}(-ri_q - L_d\omega i_d) + \frac{1}{L_q}u_q, \qquad (1.6)$$

$$M_{el} = \frac{3}{2}p(L_d - L_q)i_d i_q$$

1.1.1.4 Nonsalient-pole synchronous motor with excitation winding

$$\frac{di_d}{dt} = \frac{1}{L_1^2}(-L_f ri_d + L_{df} r_f i_f + L_f L\omega i_q) + \frac{L_f}{L_1^2}u_d - \frac{L_{df}}{L_1^2}u_f,$$

$$\frac{di_q}{dt} = \frac{1}{L}(-ri_q - L\omega i_d - L_{df}\omega i_f) + \frac{1}{L}u_q,$$

$$\frac{di_f}{dt} = \frac{1}{L_1^2}(-Lr_f i_f + L_{df} ri_d - L_{df} L_q\omega i_q) + \frac{L}{L_1^2}u_f - \frac{L_{df}}{L_1^2}u_d, \qquad (1.7)$$

$$M_{el} = \frac{3}{2}pL_{df} i_f i_q$$

where $L_d = L_q = L$ are the inductances of the stator windings.

1.1.1.5 Permanent magnet nonsalient-pole synchronous motor

$$\frac{di_d}{dt} = \frac{1}{L}(-ri_d + L\omega i_q) + \frac{1}{L}u_d,$$

$$\frac{di_q}{dt} = \frac{1}{L}(-ri_q - L\omega i_d - \Psi_f\omega) + \frac{1}{L}u_q, \qquad (1.8)$$

$$M_{el} = \frac{3}{2}p\Psi_f i_q$$

1.1.2 Semiconductor power converters

To ensure the required operating mode of the synchronous motor, the corresponding values of feed voltages should be generated on its windings. However, as the source of power a supply-line with three-phase voltages (A, B, C) with fixed amplitude and

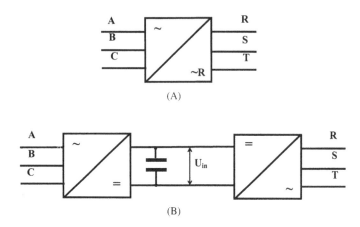

Figure 1.3 Power converter block schemas. A) Direct transformation. B) With a direct current line

frequency usually acts. Nowadays, to transform these voltages into three-phase supply voltages for a synchronous motor (R, S, T) having a variable amplitude and frequency three phase power converters are used, built using power transistors working in relay mode (Benda, 1996), (Bose, 2002), (Mohan & Underland, 2003).

The phase output voltage of the power converter represents in this case a sequence of voltage impulses. This high frequency sequence is averaged in the synchronous motor winding owing to its filtering property. The theoretical background is based on Kotelnikov's or the sampling theorem (Kotelnikov, 1933), (Mark, 1991). This average voltage could be considered, as a continuous phase voltage on the synchronous motor winding and it is a needed control to solve the control problem. Thus, the phase output voltage of the power converter from the control viewpoint has a dual character:

– Pulse – by nature;
– Continuous average – to control the electric motor.

To realize the specified transformation of the supply line three-phase voltages, the scheme of direct transformation of electric power (figure 1.3.A), or the scheme of power transformation with a direct current link (figure 1.3.B) are used.

In the first case, the power converter directly produces the three-phase voltages needed for the synchronous motor control from the three-phase supply-line voltages. The power converter that performs such transformation is named a matrix converter.

In the second case the three-phase supply-line voltages is rectified by means of a three-phase rectifier and smoothed out by an output capacitor. The constant voltage obtained arrives on the input of the three-phase voltage source inverter (VSI). From this DC line voltage the inverter produces the three-phase voltages needed for the synchronous motor control.

Let us consider these two basic types of power converters from a control viewpoint.

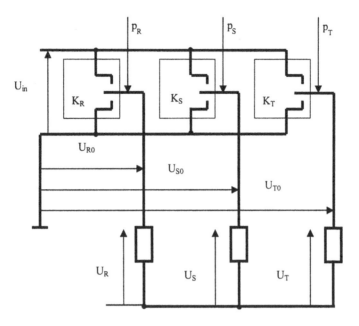

Figure 1.4 The simplified schema of the three-phase VSI

The voltage source inverter (Bose, 2002), (Mohan & Underland, 2003) transforms a DC input voltage into three-phase variable voltages of a constant or variable frequency and/or amplitude. To obtain the needed phase voltage, the voltage transformation pulse method is used. It is based on the application of a switch operation mode of power semiconductor devices (switches). They connect any output phase load to a source of constant voltage U_{in}.

One of the most widespread VSI schemes is the three-phase bridge scheme with the isolated neutral, shown in figure 1.4. It represents a parallel connection of three phases on-off power switches K_j ($j = R, S, T$). Depending on a control signal p_j ($p_j \in \{0, 1\}$) each of them connects the VSI output phase load either to a positive potential, or to the negative one of the constant input voltage U_{in}.

Thus, depending on a control signal each output phase voltage U_{j0} is equal at any moment or to input constant voltage U_{in}, or zero.

In this case the vector of output voltage VSI $U^T = (U_\alpha, U_\beta)$, in a fixed orthogonal coordinate system (α, β), is defined as:

$$\begin{vmatrix} U_\alpha \\ U_\beta \end{vmatrix} = \frac{2}{3} \begin{vmatrix} 1 & -1/2 & -1/2 \\ 0 & \sqrt{3}/2 & -\sqrt{3}/2 \end{vmatrix} \begin{vmatrix} U_{RO} \\ U_{SO} \\ U_{TO} \end{vmatrix} \qquad (1.9)$$

where numerical factors of a transformation matrix are directing the phase load orts (R, S, T). As phase voltages have relay character the voltage output vector U can accept only seven values (figure 1.5). one of which is zero U_0, and other six U_i ($i = 1, \ldots, 6$)

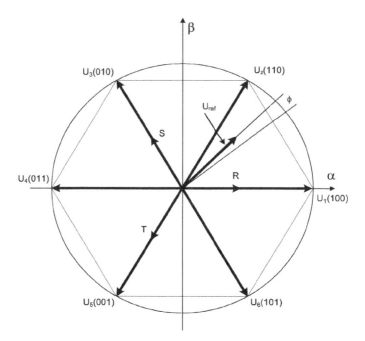

Figure 1.5 VSI momentary output vectors

are not zero and also are the tops of the correct hexagon. Its symmetry axes are directed on the phase load orts.

The module of the above mentioned six nonzero vectors depends on the load connection circuit and the value of the input dc line voltage U_{in}. In the load, connected as "star" circuit, it is equal to $2U_{in}/3$. The combinations of phase switches (p_R, p_S, p_T) controls are presented near the possible vectors of the output voltage U_i in figure 1.5. The zero vector corresponds to two combinations of the switch positions: either all are connected to positive potential (111), or to negative potential (000).

The VSI switch control, providing reception of the needed output voltage, includes two independent parts (Zinoviev, 2005), (Ryvkin & Izosimov, 1997):

– The modulation law that defines the part of the modulation period, in which the power switch are connected either to the positive potential (U_{in}) or the negative one (0);
– The switching law that defines a sequence of switching of phase switches on the modulation period.

In this case, each phase output voltage represents a sequence of squared impulses of various durations, whose amplitude is equal to U_{in}, i.e. the feed voltage. A sequence of these impulses, being averaged owing to filtering properties of the load, forms a phase output voltage on the load, which is a control tool. Thus, it is necessary to consider the dual character of an output voltage vector by analysis of the VSI work

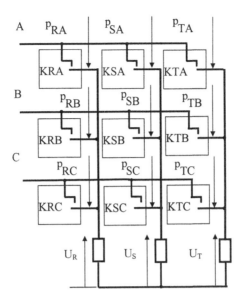

Figure 1.6 Simplified schematic circuit of matrix converter

and control design. On one hand, it is characterized at any moment by the momentary value, caused by momentary positions of the power switches. On the other hand, due to the averaging properties of the load, an average voltage value defines the VSI feature as a component of the automatic control system. These problems will be also considered below by the control design of the synchronous drive.

Matrix converters (Bose, 2002) (Mohan, 2003). Progress in the area of high-voltage and frequency operated power devices opens new possibilities for the matrix converter design. Primary benefits of such converters, which directly transform supply-line three-phase voltage (A, B, C) with constant frequency and amplitude into three-phase voltage (R, S, T) with desirable frequency and amplitude are:

- Absence of reactive elements in the basic power scheme, i.e. the elements which filter high-frequency components of the output voltage and, hence, absence of dynamic constrains;
- Any direction of a power exchange between the load and the line.

The matrix converter demerits are the raised number of operated power devices (9 or 18, depending on the applied switch scheme) and rigid requirements to the switching process. Either line short circuits or current disconnections of the load circuit during switching are inadmissible.

The schematic circuit of the matrix converter is shown in figure 1.6. The switches KRA-KTC position depends on a control signal p_{ij} ($p_{ij} \in \{0, 1\}$, $i = R, S, T$; $j = A, B, C$). The switches connect corresponding load phase (R, S, T) to corresponding supply-line phases (A, B, C). In this case the output voltage vector $U^T = (U_\alpha, U_\beta)$

in the fixed orthogonal coordinate system (α, β) is defined as:

$$\begin{vmatrix} U_\alpha \\ U_\beta \end{vmatrix} = \frac{2}{3} \begin{vmatrix} 1 & -1/2 & -1/2 \\ 0 & \sqrt{3}/2 & -\sqrt{3}/2 \end{vmatrix} \begin{vmatrix} U_R \\ U_S \\ U_T \end{vmatrix} \tag{1.10}$$

where numerical coefficients of a transformation matrix are directing the load phase orts (R, S, T). The momentary output phase voltages of the matrix converter are equal to the corresponding supply line phase voltage. The sum of these phase voltages is equal to zero, so that the three-phase condition is satisfied. The supply-line phase voltages form the three-phase system are:

$$U_A = U \sin \omega t, \quad U_B = U \sin\left(\omega t - \frac{2\pi}{3}\right), \quad U_C = U \sin\left(\omega t + \frac{2\pi}{3}\right) \tag{1.11}$$

That is why it is possible at any moment to allocate a maximum (M), an intermediate (k) and a minimum (m) values of voltage among them. They will be designated accordingly U_M, U_k or U_m. Since a primary goal of the matrix converter is the formation of the demanded output there-phase voltage, it is more expedient to describe how it works, classifying three input voltages (1.11) not on a phase accessory, but on the size of the input voltage, i.e. using the above mentioned maximum U_M, intermediate U_k and minimum U_m voltage. We will enter a load phase control p_i. This control switches three switches of a corresponding load phase and connects the load phase to one of the supply-line three phases depending on its voltage value. The load phase control p_i can accept one of three values, $p_i \in \{m, i, M\}$. If $p_i = m$ the load phase is connected to a supply-line phase with the minimum value of the voltage, if $p_i = k$ one is connected to a supply-line phase with intermediate value of voltage, and if $p_i = M$ one is connected to a supply-line phase with the maximum value of the voltage. At any moment each of load phases is connected to one of the supply-line phases. The load phase control of each load phase p_i could be transformed to the switch controls p_{ij} by using the information about the phase voltage value. As phase load control has a relay character, the output voltage vector U can accept 25 values (figure 1.7): one of which is zero U_0, and the others are 24 ones U_l $(l = 1, \ldots, 24)$ are non-zero.

Figure 1.7 shows the momentary output voltage vector diagram of the matrix converter containing three diagrams of the output voltage vectors that are typical for the VSI. The tops of the six-vector set are the tops of the correct hexagons, whose symmetry axes are directed on the directing load phase orts. The hexagon size, i.e. the value of the vector module, depends on which pair of the three input voltage values is being used, according to the momentary output three-phase voltage vectors and the time.

The above mentioned hexagons are obtained when connected to a load with only two of the three available input voltages. So the first hexagon is formed by vectors $U_1, \ldots, U_6$, where

$$U_1 = U^1, \quad U_2 = U^1 e^{j\pi/3}, \quad U_3 = U^1 e^{j2\pi/3},$$
$$U_4 = U^1 e^{j\pi}, \quad U_5 = U^1 e^{j4\pi/3}, \quad U_6 = U^1 e^{j5\pi/3}$$

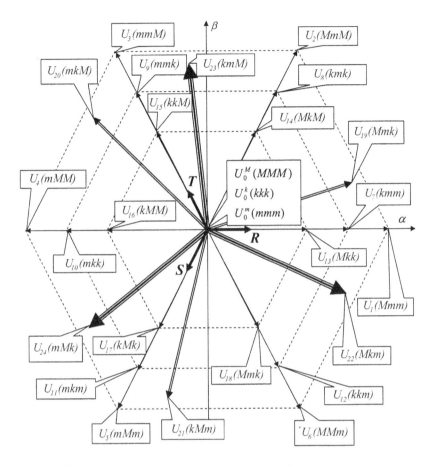

Figure 1.7 Momentary output voltage vectors of matrix converter

These vectors are obtained by connecting only the maximum U_M and minimum U_m input voltages to the load. The amplitude of the load voltage vector U^1 is not constant but varies with the sextuple of the supply line frequency in the range from $1.5U$ to $\sqrt{3}U$.

The second hexagon is formed by vectors $U_7, \ldots, U_{12}$, where

$$U_7 = U^2, \quad U_8 = U^2 e^{j\pi/3}, \quad U_9 = U^2 e^{j2\pi/3}, \quad U_{10} = U^2 e^{j\pi},$$

$$U_{11} = U^2 e^{j4\pi/3}, \quad U_{12} = U^2 e^{j5\pi/3}$$

These vectors are obtained by connecting to the load the intermediate U_k and minimum U_m input voltages only. The load voltage vector amplitude U^2 is not a constant but changes with the triple of the supply-line frequency in a range from $1.5U$ to 0.

The third hexagon is formed by vectors $U_{13}, \ldots, U_{18}$, where

$$U_{13} = U^3, \;\; U_{14} = U^3 e^{j\pi/3}, \;\; U_{15} = U^3 e^{j2\pi/3},$$
$$U_{16} = U^3 e^{j\pi}, \;\; U_{17} = U^3 e^{j4\pi/3}, \;\; U_{18} = U^3 e^{j5\pi/3}$$

These vectors are obtained by connecting to the load only the maximum U_M and intermediate U_k input voltages. The load voltage vector amplitude U^3 is not a constant but changes with the triple of the supply-line frequency in a range from $1.5U$ to 0. The change of voltage vector amplitudes of the second and third hexagons occurs in anti-phase, i.e. if the voltage vector amplitude of one of them increases the amplitude of another one decreases. The maximum moments of one of them coincide with the minimum moments of another.

Besides, there are six vectors $U_{19}, \ldots, U_{24}$ of the constant amplitude $U^4 = 1.5U$, forming two three phase systems (U_{19}, U_{20}, U_{21}), where $U_{19} = U^4 e^{j\omega t}, U_{20} = U^4 e^{j(\omega t - 2\pi/3)}, U_{21} = U^4 e^{j(\omega t + 2\pi/3)}$, and (U_{22}, U_{23}, U_{24}), where $U_{22} = U^4 e^{-j\omega t}, U_{23} = U^4 e^{-j(\omega t - 2\pi/3)}, U_{24} = U^4 e^{-j(\omega t + 2\pi/3)}$.

These two three-phase systems form a positive phase sequence and a negative one. These systems rotate with the supply-line frequency and correspond to possible connections of the three-phase load to a three-phase supply-line. Figure 1.7 presents the possible vectors of the output voltage U that correspond to the combinations of phase load controls (p_R, p_S, p_T). A zero vector U_0 corresponds to three combinations of the phase load controls: all three load phases are connected or to maximum input phase voltage (MMM), or to intermediate one (kkk), or to minimum (mmm) one. A zero vector will have the top index value of the used voltages U_0^M, U_0^k, U_0^m corresponding to each of these cases.

The diagram of current vectors of the supply-line is similar to the diagram of load voltage vectors. The differences are the following: instead of the load orts and the maximum, intermediate and minimum values of the supply-line voltages the supply-line ones and ones of load phase currents are respectively used. The current consumed from a supply-line has the impulse form; the amplitude being equal to a load current. For current smoothing and decreasing interferences created by the matrix converter a supply-line filter could be applied. The needed values of the reactive filter elements of this filter are essentially less than the values of a choking inductance and capacity of a VSI DC-link, considered in the previous section.

It is possible, using modulation, to obtain the needed average values of the output voltages or the input currents.

Let us notice that there is another possibility to use the matrix switch scheme, similar to the matrix converter scheme to improve the quality of the converter power consumption. The switch scheme could connect a line-supply and a DC-link. Thus, if the choking inductance is established at the input of such scheme, it is equivalent to the voltage source inverter considered earlier, and if the choking inductance is established at the output, such schema is equal to the current inverter, respectively.

1.2 DRIVE CONTROL PROBLEMS AND THEIR EXISTING SOLUTIONS

An electrical motor in the drive structure carries out a transformation of the electric energy into a mechanical one that moves the mechanisms participating in the working process. The technology requirements for this process define the necessity and expediency of maintenance at the reference level of those or other mechanical variables, e.g.: position, speed, acceleration, torque etc, of the mechanism tip (Leonhard, 2001), (Boldea & Nasar, 2005), (Bose, 2002), (Il'insky, 2003).

The main adjustable variable is usually a mechanical coordinate, more often a rotor speed $\Omega(t)$ or a rotor position $\Gamma(t)$, which should be equal to reference values $\Omega_z(t)$ ($\Gamma_z(t)$). The reference value $\Omega_z(t)$, in the most general case, is a time function. That is essentially the servo control problem. The actual rotor angular speed should reproduce all changes of the reference with accuracy. Special cases of this problem are:

– Stabilization of the rotor speed at the reference constant level;
– The rotor angular speed change under the reference law;
– Restricting the angular speed by an admissible value.

In relation to working off the reference demands, any change of the load torque has to be carried out exactly and fast.

Together with the requirements on the mechanical coordinate control, demands on profitability of drive work are made. Its indicators are the efficiency coefficient, characterizing power losses, and $\cos\varphi$, characterizing consumption of the reactive power.

Thus, the control goal is maintaining the reference value of the rotor velocity $\Omega(t)$ in combination to power requirements performance. Since the real physical controls are switch converter controls, defining a value and a sign of discontinuous (relay) control, the control problem consists in the formation of such switch controls so that the above mentioned requirements are fulfilled. Therefore, referring to the theory of systems with discontinuous controls is quite natural in this case.

Essential nonlinearities of both power converters and synchronous motors and the complication of such drives bring up a variety of approaches to the solution of a control design problem.

We can set apart several basic approaches to the design of synchronous drive controls (Leonhard, 2001), (Pahman & Zhou, 2000), (Bose, 2002), (Il'insky, 2003), for instance:

– One-loop control;
– Decomposition one-loop control;
– Cascade (subordinated) control.

In the first case the drive is considered as a unit. The control design is interfaced to the solution of the difficult nonlinear problem. Its complexity is caused by essential nonlinearities of the drive elements. The switching frequency and duration of switching of various power switches are generated automatically in the closed loop, as an auxiliary element in the solution of the primary goal of drive control. Such dynamic systems possess high dynamics and small sensitivity to drive parameter changes and external disturbances. Unfortunately, the automatically generated switching frequency of the power switches in this case is not a constant; it depends on the initial conditions. This leads to power switches losses increase and to drive mechanical noise.

In the second case two independent problems are considered, namely, the control design of a synchronous motor and that of a power converter. The control for the synchronous motor is synthesized on the assumption that the power converter generates the synchronous motor input voltages responding to the solution of the primary goal of control. The problem of the power converter control consists in transforming the constant input voltage of an alternating or a direct current in three-phase voltage with variable amplitude and frequency providing the general solution of the control problem.

This power converter control problem deals independently with the use of the modulation based on a high-frequency connection to the input voltages of the load phases. In this case each phase output voltage of the power converter represents a sequence of impulses of various durations, whose amplitude depends on the amplitude of a momentary used input voltage. The sequence of these impulses, being averaged owing to the filtering feature of the load, forms a continuous phase output voltage on load, which is the motor control. In this case the indicators characterizing the discontinuous control, i.e. its modulation, such as switching frequency and pulse ration, are formed for the power converter from the outside, i.e. feed forward (program) control takes place. However, the automatic compensation of external disturbances and internal parameter changes using such control is possible only in the closed loop of the mechanical coordinate control.

The (subordinated) control approach is used by control design in the third case. It is based on the decomposition of an initial problem of the drive control on processes rates in the drive. The control problem for each of processes is handled independently. The control problem in the drive is divided by natural splitting of the processes: the electromagnetic ones are fast and the mechanically ones are slow. Moreover, in relation to fast processes, mechanical variables are quasi constants. Hence the electromagnetic processes control problem is reduced to a control of the electromagnetic torque. The torque reference is formed in a slow control loop of the mechanical movement, which is based on (1.1). Since the value of the electromagnetic torque is defined by the values the currents in the synchronous motor windings, the control problem for a fast internal loop is reduced to a feedback current control using the power converter. The character of change of discontinuous controls, i.e. the switching frequency and duration of the switching of various power switches are generated automatically in the closed loop, as an auxiliary element of the solution of a current control problem. The current control loop possesses high dynamics and small sensitivity to motor parameter changes and external disturbances. However, it is necessary to notice that, as well as in the first case, the switching frequency of power switches is generated automatically and its value depends on initial conditions. It can lead to an increase of switching losses by the power switches and mechanical drive noise.

Design problems of a high-quality drive are inseparably linked to problems of receiving the information of process state variables, in particular, about controlled coordinates.

The traditionally used approach based on direct measurement of all the necessary coordinates, leads to considerable complication of the drive design, its operational deterioration and cost indexes. A possible way of overcoming these demerits is the exclusion of drive sensors for those coordinates whose direct measurement is undesirable, and the control design using a state observer to obtain estimates (Leonhard, 2001), (Boldea & Nasar,), (Andreescu et al, 2000), (Dote, 1988).

The total elimination of mechanical coordinate's sensors and design of a sensorless drive, containing only sensors of electric variables, is in perspective. The complexity of such approach is caused by the electromagnetic part of an electric motor, and in particular the synchronous motor, as shown in section 1.1.1. It is described by the nonlinear equations (1.4)–(1.8). Working out new nonlinear methods of estimation of mechanical coordinates is necessary for reception on electric variables as estimations of mechanical coordinates.

Features of control and observer design based on the theory of systems with discontinuous controls for the drives used, the above presented types of synchronous motors, and power converters will be considered in subsequent chapters. The new approaches to drive control and observer design in continuous and discrete time, and designed with the help of novel algorithms to improve control quality are presented.

REFERENCES

Andreescu G.D., Popa A. and Spilca A. "*Sliding mode based observer for sensorless control of PMSM drives – two comparative study cases*". Proc. the 7th International Conference on Optimization of Electrical and Electronical Equipment, OPTIM 2000, Brasov, Romania, 2000, CD-ROM.

Benda V. "*Reliability of power semiconductor devices – Problems and trends*". Proc. 7th International Power Electronics & Motion Control Conference, PEMC'96, Budapest, Hungary, 1996, vol. 1, pp. 30–35.

Boldea I. and Nasar S.A. "*Electric drives*", 2nd ed. CRC Press, 2005. 544 p.

Bose B.K. "*Modern power electronics and AC drives*". New Jersey: Prentice Hall, 2002. 711 p.

Consoli A. "*Advanced control techniques. Modern Electrical Drives*". Dordrecht, Boston, London: Kluwer Academic Publishers, 2000, pp. 523–582.

Dote Y. "*Application of modern control techniques to motor control*". Proc. of the IEEE, 1988, vol. 76, no. 4, pp. 438–445.

Hanitsch R. and Parspour N. "*Exterior Permanent Magnet Motors. Modern Electrical Drives*". Dordrecht, Boston, London: Kluwer Academic Publishers, 2000, pp. 79–114.

Holtz J. "*Sensorless control of induction machines – with or without signal injection*". Proc. the 9th International Conference on Optimization of Electrical and Electronical Equipment, OPTIM 2004, Brasov, Romania, 2004, vol. II, pp. XVII–XXXIX.

Il'insky N. "*Drive backgrounds*". Moscow: Publishing house of Moscow power engineering institute, 2003. 221 p. (in Russian).

Kotelnikov V.A. "*On the carrying capacity of the ether and wire in telecommunications*", Material for the First All-Union Conference on Questions of Communication, Izd. Red. Upr. Svyazi RKKA, Moscow, 1933. (in Russian).

Leonhard W. "*Control of electrical drives*". Berlin: Springer-Verlag, 2001. 460 p.

Marks R.J. II: "*Introduction to Shannon sampling and interpolation theory*". New York: Spinger-Verlag, 1991. 332 p.

Mohan N., Underland T.M. and Robbins W.P. *"Power electronics: converters, applications and design"*. 3rd edition. New York: John Wiley & Son Inc., 2003. 824 p.

Pahman M.A. and Zhou P. *"Interior permanent magnet motors. Modern Electrical Drives"*. Dordrecht, Boston, London: Kluwer Academic Publishers, 2000, pp. 115–140.

Park R. *"Two-reaction theory of synchronous machines"*. AIEE Transactions, 1929, vol. 48, pp. 716–730, 1933, vol. 52, pp. 352–355.

Ryvkin S.E. and Izosimov D.B. *"Comparison of pulse-width modulation algorithms for three-phase voltage inverters"*. Electrical Technology, 1997, no. 2, pp. 133–144.

Zinoviev G.S. *"Power electronics backgrounds"*. Novosibirsk: Publishing house of Novosibirsk state technical university, 2005. 664 p.

Chapter 2

Sliding mode in nonlinear dynamic systems

2.1 PLANT FEATURES AND SLIDING MODE DESIGN

The considered synchronous electrical drive, based on a complex semiconductor power converter and a synchronous motor, from a control viewpoint, represents a nonlinear dynamic system with a linear control input signal $u(t)$, whose discontinuous character is caused by a switching operation mode of the elements of the semiconductor power converter.

$$\frac{dx(t)}{dt} = f(x,t) + B(x,t)u(t) \tag{2.1}$$

where $x(t)$ is a state vector, $x(t) \in R^n$; $f(x,t)$ is a column-vector of the control plant, $f(x,t) \in R^n$; $u(t) \in R^m$; $B(x,t)$ is a matrix before the control, having dimensions $n \times m$.

Prominent features of the considered class of nonlinear dynamic systems, along with discontinuous character of controls, are specified below:

- The discontinuous controls $\tilde{u}_i(t)$, which undergo ruptures on the surfaces $S_i(x) = 0$ in the system state space, can be modeled like this:

$$\tilde{u}_i(x,t) = \begin{cases} \tilde{u}_i^+(x,t) & if \ S_i(x,t) > 0 \\ \tilde{u}_i^-(x,t) & if \ S_i(x,t) < 0 \end{cases} \tag{2.2}$$

- The discontinuous control number $i = \overline{1,q}$ outnumbers a dimension of the control space U, $\dim U = m$; $q \geq m$;
- There is a set of fixed orts in the control space $\tilde{E} = \{e_i\}$, $i = \overline{1,q}$ $q \geq m$ defining directions of the discontinuous component of a control vector $\tilde{u}^T(t) = [\tilde{u}_1(t), \ldots, \tilde{u}_q(t)]$. Only the so directed components of the control vector should be used for the control design;
- matrix $B(\mathbf{x},t)$ is periodic and has a period T

$$B(x,t) = B(x,t+T) \tag{2.3}$$

The discontinuous character of the controls does a natural reference to solve a design problem of electrical drive control in the theory of systems with sliding mode.

The dynamic nonlinear systems with the sliding mode control possess very attractive properties, such as high quality of control, invariance to external perturbations, tolerance to changes of plant parameters (Emelyanov et al, 1970), (Utkin, 1978), (Utkin, 1992), (Utkin et al, 1999), (Zinober, 1994), (Edwards & Spurgeon, 1998).

However the prominent features of system listed above do not suppose direct use of known results of the theory of systems with sliding movements because known results of this theory are received, first of all, in the assumption that discontinuous controls $\tilde{u}_i(x, t)$ form the basis of a control vector $u(t)$, i.e. the discontinuous control number is equal to the dimension of the control space, $q = m$.

It is therefore necessary to formulate the existence conditions of sliding mode in the systems under research and to develop methods for the control design taking into account the above mentioned features, the peculiar features of various control schemes of three-phase electrical drive, and the various types used in them three-phase semiconductor power converters and synchronous motors.

The suggested approach to solve the above mentioned problems is based on breaking down an initial control problem into more simple ones using their step solution. In the first step, the sliding motion design in system (2.1) with a constant square matrix of a full rank B and discontinuous controls, forming the basis in the control space U, is carried out. The design result is areas of admissible controls, ensuring the existence of a sliding mode. In the second step, the coincidence problem for the areas of admissible and realized controls according to (2.3) is solved.

The projection of a discontinuous control vector $\tilde{u}(t)$ to the control space U is calculated according to the following transformation:

$$u = M(t)\tilde{u} \tag{2.4}$$

where M is a matrix of projection of discontinuous control vector on control space, whose dimension is $m \times q$.

Let us notice that in this case at each time in control space U there is a set of realized control vectors $\tilde{U} = \{U_r\}, r = \overline{1, 2^q}$.

It is known that in systems with discontinuous controls this specific kind of movement called sliding mode could appear. It is widely used to solve control problems (Emelyanov et al, 1970), (Utkin, 1978), (Utkin, 1992), (Utkin et al, 1999), (Zinober, 1994), (Edwards & Spurgeon, 1998). As it was marked above, such kind of motion possesses a number of attractive properties: high quality of control, possibility of invariance to external unmeasured influences, small sensitivity to changes of dynamic properties of the plant, and so on.

The basic question is under what conditions sliding motion exists in a discontinuous system. The problem of a sliding mode existence is equivalent to a stability problem of an initial system (2.1)–(2.4). It is solved in a l-dimensional error space of the variables $Z(t) = z_z(t) - z(t)$, where $z(x, t)$ is the control variables vector, $z(x) \in R^l$, and $z_z(x, t)$ is a reference of the control variables vector. In the theory of systems with discontinuous controls, zero errors of the components of control variables $Z_j(x, t) = 0$, $j = \overline{1, l}$, or their linear transformations, are also named switching surfaces, since the signs of discontinuous controls change on them. Using the terminology of the stability theory for nonlinear systems, it is possible to speak about existence conditions of a sliding mode "in small" and "in big". The concept stability "in small" is equivalent to a condition of existence of sliding movement on a switching surface or on crossing of such surfaces. Also, the concept of stability "in big" along with existence of sliding movement on a switching surface or on crossing of such surfaces defines and "the reaching condition", i.e. a condition, which performance guarantees that a representing point from any initial position reaches a surface or crossing of surfaces of switching.

For the problem solving of the existence of sliding movement usually the method of stability definition of Lyapunov (Emelyanov et al, 1970), (Utkin, 1978), (Utkin, 1992), (Utkin et al, 1999), (Zinober, 1994), (Edwards & Spurgeon, 1998) is used. In this case, the equations of a projection of movement of initial system (2.1)–(2.4) on a subspace of the control variables errors are analyzed:

$$\frac{dZ}{dt} = \frac{dz_z}{dt} - (Gf + D\tilde{u}) \tag{2.5}$$

where G is a matrix of the dimension $l \times n$, which line are vectors-gradients of functions $Z_j(x, t)$; $D = GBM$.

Lemma: The dimension of the control variables vector $z(t)$ cannot outnumber a rank of any factor matrixes of matrix D, i.e. $l \leq \min[rank\ G,\ rank\ B(x,t),\ rank\ M]$.

In order to provide the control problem solving by using a classical sliding mode in system (2.5) it is necessary that the dimension of an control variable error space or that the same one of the control variable space would be less or equal to a projection of the control space on this above mentioned ones. The projection dimension is defined by a matrix D rank. According to a consequence from the Binet-Cauchy formula the rank of this matrix cannot outnumbers the minimum rank of the factor matrixes. Thus, the maximum legitimate dimension of a control variable vector $z(t)$ is equal to a matrix D rank, i.e. $l \leq \min [rank\ G,\ rank\ B(x,t),\ rank\ M]$. The statement is proved.

Let us notice that fact about the ranks of the factor matrixes. The matrix of gradients G gets out proceeding from a control problem in view and its rank, at desire, can be set as much as possible achievable, $rank\ M = m$, but the matrix $B(x, t)$ rank is set a priori, proceeding from physical nature of the plant.

We already have a "classical" statement of a control problem in discontinuous systems with sliding movement, when:

- The dimension of the control space is less or equal to the dimension of the state space $m \leq n$;
- Matrix $GB(x, t)$ is a matrix of a full rank, $rank\ GB = l$;
- The dimensions of the control variable vector $z(t)$ and of the control one $u(t)$ are equal, $l = m$;
- The discontinuous controls $\tilde{u}_i(x, t)$ are components of the control vector $u(t)$, $q = m$.

There is a variety of sufficient existence conditions for the existence of sliding mode for the above mentioned classical systems. Various methods for the design of this movement were developed based on them. Let us briefly consider those used in this work.

As it is known (Emelyanov et al, 1970), (Utkin, 1978), (Utkin, 1992), (Utkin et al, 1999), (Zinober, 1994), (Edwards & Spurgeon, 1998) by the scalar control an existence condition of a sliding mode on the switching or sliding surface $Z(x) = 0$ is

$$\lim_{Z \to +0} \frac{dZ}{dt} < 0, \quad \lim_{Z \to -0} \frac{dZ}{dt} > 0 \tag{2.6}$$

It must be emphasized that it is a necessary and a sufficient condition, and it means that a deviation from the sliding surface and the speed of its change, i.e. its time derivative, have opposite signs in the neighborhood of this sliding surface.

There is no universal existence condition for a sliding mode on crossing switching surfaces when dealing with vectors. The majority of known existence conditions of multidimensional sliding movement are closely connected to a condition of stability concerning variety $Z(x) = 0$, and just using for the control problem solving the second stability method of Lyapunov. The received sufficient conditions are formulated with reference to a matrix D of the equation (2.5) (Utkin, 1978), (Utkin, 1992), (Utkin et al, 1999), (Zinober, 1994), (Edwards & Spurgeon, 1998). Some of them used in this book are presented below.

2.1.1   Sufficient existence conditions of a sliding mode

In system (2.1)–(2.4) the sliding mode on manifold $Z(x) = 0$ exists, if one of following conditions is satisfied:

a) Matrix D is one with a predominant diagonal, i.e. $|d_{ii}| > \sum_{\substack{j=1 \\ i \neq j}}^{m} |d_{ij}|$ (d_{ij}, $i = \overline{1, m}$, $j = \overline{1, m}$ is a matrix D element), and discontinuous control is chosen as

$$u_i(x, t) = \begin{cases} -M_i(x, t) & \text{if } Z_i d_{ii} > 0 \\ M_i(x, t) & \text{if } Z_i d_{ii} < 0 \end{cases} \tag{2.7}$$

with amplitude

$$M_i(x, t) > \frac{|q_i| + \sum_{\substack{j=1 \\ j \neq i}}^{m} |d_{ij}|}{|d_{ii}|} \tag{2.8}$$

where $q_i(x, t)$ is a vector Gf element.

In this case, on each of surfaces $Z_i(x) = 0$ the sliding mode existence condition is satisfied, i.e. the m-dimensional sliding mode breaks up on m a one-dimensional ones.

b) There is a hierarchy of controls in the system that allows reducing multidimensional control problems to consistently solving m one-dimensional ones. In this case, one of the vector control components, e.g. u_1 provides movement a sliding mode on a surface $Z_1(x) = 0$ irrespectively of other control values. After appearance of sliding movement on the surface $Z_1(x) = 0$, control u_2 provides movement on crossing of the sliding surfaces $Z_1(x) = 0$ and $Z_2(x) = 0$, irrespectively of values of the others of control, etc. Using such approach of sufficient existence conditions of multidimensional sliding mode, the solution is based on similar conditions for a scalar case (2.6)

$$gradZ_{k+1}b_k^{k+1}u_{k+1}^+ < \min_{u_{k+2},\dots,u_m} \left[-gradZ_{k+1}f^k - \sum_{j=2}^{m-k} gradZ_{k+1}b_k^{k+j}u_{k+j} \right],$$

$$gradZ_{k+1}b_k^{k+1}u_{k+1}^- > \max_{u_{k+2},\dots,u_m} \left[-gradZ_{k+1}f^k - \sum_{j=2}^{m-k} gradZ_{k+1}b_k^{k+j}u_{k+j} \right]$$

(2.9)

where k is a number of the switching surfaces on which there is a sliding mode, $0 \le k \le m - 1$; f^k is a n-dimensional vector, B_k is a matrix of dimension $n \times (m - k)$ with columns $b_k^{k+1}, \dots, b_k^m$; f^k and B_k are elements of the differential equation describing initial dynamic system (2.1)–(2.4) in the presence of a sliding mode on crossing of k switching surfaces.

It is obvious that the sufficient existence condition of the sliding mode, received with use of a method of control hierarchy, coincides with the sufficient condition received in the case (a), when a matrix D is diagonal.

For the description of system (2.1)–(2.4) sliding movement on all manifolds $Z(x) = 0$ or on its part that is necessary by using of the hierarchy method the method of equivalent control (Utkin, 1978), (Utkin, 1992), (Utkin et al, 1999) is used. It can be proved that movement in sliding mode can be described with use of continuous control that is called as equivalent control u_{eq}. This control ensures the time derivative to be zero in a vector $Z(x)$ on the system trajectories:

$$u_{eq} = (D)^{-1}Gf \tag{2.10}$$

The received equivalent control u_{eq} is substituted in the initial system of the equations (2.1) and the resulting equation describes the system movement in sliding mode:

$$\frac{dx(t)}{dt} = f(x,t) + B(D)^{-1}Gf \tag{2.11}$$

The motion equations in a sliding mode (2.11) for a given system structure depend only on the elements of matrix G. Hence, after changing position of the switching surfaces in the system state space, it is possible to influence the nature of motion in sliding mode. In addition, the problem of designing the desirable motion in a sliding mode is a problem of lower order than the original. It is because the motion in sliding mode along with the differential equation (2.11) describes more m algebraic equations switching surfaces $Z(x) = 0$, which allows lowering the order of equation (2.11) for m.

Thus, the control design problem in systems with discontinuous controls generally breaks up into three problems:

-   Design problem of the motion in a sliding mode;
-   Sliding mode existence problem;
-   Reaching problem on sliding manifold.

The first problem deals with a choice of corresponding functions of the switching providing desirable motion in the sliding mode. In this case, the classical design approaches of the automatic control theory for this problem solving can be used since the right part of the differential equations of sliding in this case is continuous.

The second and third problems become complicated sufficient character of existence conditions of sliding motion. Moreover, due to the first sufficient condition it is necessary to transform matrix D to a matrix of special kind. The transformation of a matrix before control in the equation (2.5) to a special kind is made not at the cost of a corresponding choice of a matrix G, which should provide the solution of the first problem, but at the expenses of use of matrixes linear non-singular transformation or a control vector $R_u(x, t)$

$$u^* = R_u(x, t)u \tag{2.12}$$

where u^* is a new control vector, or the switching surfaces $R_S(x, t)$

$$Z^* = R_S(x, t)Z \tag{2.13}$$

where Z^* is a vector of new switching surfaces. It is possible, thanks to invariance property of the sliding equations to accomplish such transformations. The design involves a choice of desirable switching surfaces. Then, such transformation of these surfaces or a control vector by using transformation matrixes (2.13) or (2.12) accordingly, is made, so that one of the known sufficient existence conditions of a sliding mode or reaching conditions could be used in control design.

It is necessary to notice that matrix D transformation using (2.12) and (2.13) into a diagonal matrix leads to various consequences.

By using a non-degenerate linear transformation of a control vector, the transformation matrix $R_u(x, t)$ gets out in a kind of the matrix D and the transformed equation of initial system becomes

$$\frac{dZ}{dt} = Gf + u \tag{2.14}$$

In this case, the received sufficient condition depending on those points x, for which they are carried out, is an existence conditions or even a reaching conditions, i.e. an existence of sliding motion both "in small", and "in big" is provided.

In case of using a linear non-degenerate transformations of the switching surfaces the transformation matrix $R_S(x, t)$ gets out in a kind of the matrix $(D)^{-1}$ and the transformed equation of initial system becomes

$$\frac{dZ^*}{dt} = (D)^{-1}Gf + u + \frac{d(D)^{-1}}{dt}DZ^* \tag{2.15}$$

Unlike the previous case the matter can be only about the existence condition since last summand of this equation is absent only in the presence of sliding motion. The question on whether the sliding manifold reaching can only be resolved if there is the additional information about the summand $(d(D)^{-1}/dt)DZ^*$.

2.2 SUFFICIENT EXISTENCE CONDITIONS OF SLIDING MODE IN SYSTEMS WITH REDUNDANT CONTROL

The presence in system (2.3) of the above specified features (namely, the periodicity of a matrix before the control vector $B(x, t + T)$, the number of discontinuous controls being larger than a dimension of control space $q > m$, and their rigid binding to directions of the ort sets $\tilde{E}$, makes the direct use of classical results of the theory of systems with discontinuous controls impossible. Therefore, it is necessary to formulate the existence conditions for a sliding mode in these systems and to develop methods for the control design taking into account the above mentioned specified features of a nonlinear system. The following are conditions of existence of sliding mode in dynamic systems with discontinuous controls and the sliding mode design methods for these systems are developed and presented below (Ryvkin, 1984).

Theorem 1: In a system with a linear control (2.1)–(2.4) there is sliding mode in the crossing of surfaces $Z(x) = 0$, $Z(x) \in R^l$, if for all $x \in Z(x) = 0$ on the period T, the following conditions are satisfied:

–   The dimension of control vector is equal to the dimension of sliding space $l = m$;
–   The matrix $D^* = GBM$ has the rank m. (G is a matrix of dimension $l \times n$, whose rows are vectors-gradients of function $Z_j(x)$).

Proof: It is always possible to assign the basis of dimension l in a matrix D^*. However columns of the matrix D^* forming it owing to periodicity of a matrix B cannot be fixed.

In this case, the period T can be divided into intervals of a constancy of basis. On each of these intervals the equation of a projection of movement of initial system (2.1)–(2.4) on a subspace of a control variable error can be written down as follows:

$$\frac{dZ}{dt} = \frac{dz_z}{dt} - (Gf + Du^* + d^*u^{**}) \tag{2.16}$$

where $D^* = |Dd^*|$, D is a matrix $l \times l$, $rank\ D = l$; d^* is a matrix $l \times (m - l)$; $\tilde{u}^T = (u^*u^{**})$, $u^* \in R^l$.

Thus, the equation member d^*u^{**} could be considered as an external influence.

Since the sliding motion equations invariance to linear transformations of switching surfaces $Z(x) = 0$ by bringing in new switching surfaces $Z^* = 0$ (2.13), the matrix before a control vector u^* can be transformed to a diagonal kind and the sufficient condition of a sliding mode existence (2.7), (2.8) is provided.

Thus, the statement is proved.

Using the above mentioned approach, the unused controls are considered as unwanted disturbances with which it is necessary to deal. It would be desirable to use available redundancy on control for control problem solving.

Theorem 2: In linear on control system (2.1)–(2.4) with periodic factors before control and redundant control there is a sliding movement on crossing of surfaces $Z(x) = 0$,

$Z(x) \in R^l$, if for all $x \in Z(x) = 0$ following conditions are satisfied:

- The dimension of control vector is equal to the dimension of sliding space, $l = m$;
- The square matrix $D = GB$ of dimension $m \times m$ is nonsingular (G is a matrix of dimension $l \times n$, whose rows are vectors-gradients of functions $Z_j(x)$);
- The intersection of realizable control vector $\tilde{U} = \{U_r\}$, $r = \overline{1, 2^q}$ with each of 2^m domains of control space to ensure the existence of sliding movement, is not empty.

Proof: Let us consider the equations of a projection of movement of initial system (2.1), (2.2) on a subspace of a control variable error

$$\frac{dZ}{dt} = \frac{dz_z}{dt} - (Gf + Du) \tag{2.17}$$

Since the sliding equations invariance to linear transformations of control, we will enter a new control vector $v(t)$, $v(t) \in R^l$, which is connected by linear transformation with an initial control vector $u(t)$

$$u = M^*(t)v \tag{2.18}$$

where $M^*(t)$ is a square matrix of transformation of dimension $m \times m$, *rank* $M^* = m$, which gets out of a condition that the matrix before the new control $v(t)$ in the equation (2.17) provided performance of sufficient existence conditions of a sliding mode, by means of one of known methods of design of sliding movement, e.g. diagonalization (2.7), (2.8), control hierarchy (2.9).

It will be assumed for descriptive reasons that due to the use of matrix $M^*(t)$, matrix $D^*(x)$ before control in the rewritten equation (2.17) is a nonsingular diagonal matrix with constant elements, i.e.

$$\frac{dZ}{dt} = \frac{dz_z}{dt} - (Gf + D^*v) \tag{2.19}$$

where $D^* = DM^*$, d_j^* is a diagonal element, $j = \overline{1, m}$, $\inf_j |d_j^*| > 0$.

It is obvious that since a matrix D is nonsingular, it is possible to choose a matrix $M^*(t)$, inverse to matrix D, i.e. $M^*(t) = D^{-1}$, and the matrix before control will be a unitary one E.

In this case, the existence problem of a multidimensional sliding mode breaks up on m one-dimensional problems, each of which dares on a basis existence conditions of a one-dimensional sliding mode (2.6).

Every component of the control variable error vector $Z_j(t)$ fits with the control vector component $v_j(t)$ provided that the control variables errors are equal to zero in sliding mode. There are m such pairs

$$v_j(t) = V_j \text{sgn}(Z_j) \tag{2.20}$$

where V_j is a control component module that value is

$$V_j \geq V_{jeq} = \left| \frac{dz_{jz}}{dt} - grad(Z_j)f \right| \tag{2.21}$$

where $grad(Z_j)$ is a matrix G vector-line; V_{jeq} is an equivalent control that is an continuous control received by equating to zero the time derivative from a vector component vector Z_j on the system trajectories.

As a result in space of new control $v(t) \in V$, dim $V = m$, the area of the admissible controls V^* is selected. It consists of 2^m areas $V_p^*, p = \overline{1, 2^m}$, each of which is defined by a combination of signs of the control vector components $v_j(t)$ (2.20) and inequalities on the value of their modules (2.21) providing sliding mode, i.e.

$$V^* = V_1^* \cup V_2^* \cup \cdots \cup V_p^* \cup \cdots \cup V_{2m}^*, \quad V_p^* \cap V_a^* = 0, \quad p \neq a, \, a = \overline{1, 2^m} \tag{2.22}$$

While formulating the problem, the definition of sliding motion on the manifold $Z = 0$ should be provided with the use of initial discontinuous controls $\tilde{u}_i(t)$ (2.3). Let us consider the original space control U, which, according with the transformation (2.4), contains a set of realizable control vectors $\tilde{U} = \{U_r\}$, $r = \overline{1, 2^q}$ obtained using the discontinuous control $\tilde{u}_i(t)$ (2.3).

Using the linear transformation (2.18) the received area of admissible controls V^* (2.22) could be projected on the same control space U. It is obvious that owing to linear control transformation the sliding movement of dynamic system with discontinuous controls (2.1)–(2.4) on variety $Z = 0$ exists, if at any moment each of projections of subareas of admissible controls V_p^* on the control space U, at least one vector from the set of realized control vectors $\tilde{U} = \{U_r\}$ belongs. Thus, the theorem is proved.

It is obvious that the concrete definition of a kind of matrix B before the control and a matrix of gradients G allows concretizing conditions of sliding movement.

Let us consider a consequence of the theorem, which is of great importance for the drive control design.

Consequence. Sliding mode in dynamic system (2.1) of dimension 2 with a periodic square-law matrix B of full dimension

$$B = \left| \begin{matrix} \cos(at) & \sin(at) \\ -\sin(at) & \cos(at) \end{matrix} \right| \tag{2.23}$$

where a is a factor, and with three discontinuous, equal on the module controls $u(t) \in R^2$ $q = 3$:

$$\tilde{u}_i^+(x, t) = -\tilde{u}_i^-(x, t) = u_0 \tag{2.24}$$

and a projection matrix

$$M = \left| \begin{matrix} 1 & -\sqrt{3}/2 & -\sqrt{3}/2 \\ 0 & 1/2 & -1/2 \end{matrix} \right| \tag{2.25}$$

exists, if the module of discontinuous controls is more than some value A.

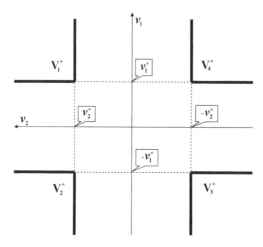

Figure 2.1 Area of admissible control $v^T(t) = (v_1, v_2)$

Proof: Let us use in this case a transformation matrix $M^*(t)$, *rank* $M^* = 2$,

$$M^* = \begin{vmatrix} \cos(at) & -\sin(at) \\ \sin(at) & \cos(at) \end{vmatrix} \tag{2.26}$$

In this case in the equation of the dynamic system a matrix before a new control vector $v(t)$, $v(t) \in R^2$ is unitary one

$$BM^* = \begin{vmatrix} 1 & 0 \\ 0 & 1 \end{vmatrix} = E \tag{2.27}$$

Evidently in this case, the state matrix of the projection in equation (2.5) will be determined by the choice of matrix G, i.e. the selected control variable vector $z(t)$, $z(t) \in R^2$.

If we assume that the control variables $z(t)$ are chosen in such a manner that a matrix D in equation (2.5) has a full dimension, the new control vector components $v^T(t) = (v_1, v_2)$ can be designed by using one of the well known design methods, i.e. diagonalization (2.7), (2.8) or control hierarchy (2.9).

The result of the sliding mode design using the discontinuous control vector $v(t)$ is a selection in the new control space $v(t) \in V$ of an area of admissible controls V^* (figure 2.1), consisting of 4 subareas V_p^*, $p = \overline{1,4}$. Each of them is defined by a combination of signs control vector components $v_j(t)$ $j = 1, 2$

$$v_j(t) = V_j \operatorname{sgn}(Z_j) \tag{2.28}$$

and inequalities in their module value

$$|v_j(t)| = V_j \geq v_j^* \tag{2.29}$$

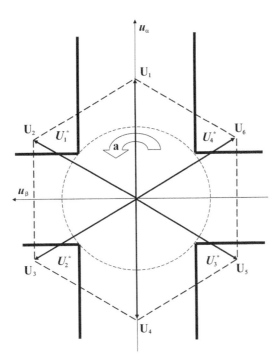

Figure 2.2 Areas of admissible and realized control $u(t)$

providing sliding mode.

$$V^* = V_1^* \cup V_2^* \cup V_3^* \cup V_4^*, \quad V_p^* \cap V_r^* = 0, \quad p \neq r, \, r = \overline{1,4} \qquad (2.30)$$

According to the problem definition, the control variable errors must be brought to zero $Z = 0$ by using an initial vector of discontinuous control $\tilde{u}(t)$ (2.3), i.e. maintenance of sliding mode on manifold $Z = 0$ should be implemented using the above mentioned initial vector of discontinuous control $\tilde{u}(t)$.

Let us consider, for convenience of analysis, all possible realizations of a control vector $u(t)$ obtained with use of three discontinuous controls $\tilde{u}_i(t)$, in the control space, and a projection to this control space, the areas of the controls V^* necessary to ensure sliding mode.

The spaces of initial $u(t)$ and new $v(t)$ controls are connected by linear transformation (2.27), i.e. the displayed area of needed controls has the same configuration, as well as initial area, but its position is not constant. It rotates with a frequency equal to the factor 'a'.

Since amplitudes of discontinuous controls are equal, the control vector $u(t)$ can accept six nonzero values $U_1, \ldots, U_6$ with the identical module $U_0 = 2u_0$, located on distance $\pi/3$ from each other (figure 2.2). At the same time the projection of the area of the needed for the sliding mode realization controls U^* is represented by a set of four symmetrically located subareas $U_1^*, \ldots, U_4^*$ rotating with frequency a.

It is possible to corroborate by direct check that on all period $T = 1/a$ every subarea of the needed control area has at least one of six realized nonzero control vectors $u(t)$, if the following condition is satisfied:

$$arctg \frac{v_1^*}{2u_0} + arctg \frac{v_2^*}{2u_0} \leq \frac{\pi}{6} \tag{2.31}$$

Resolving inequality (2.31) it is fair to say that

$$u_0 \geq u_0^* = \frac{1}{2}\sqrt{(v_1^*)^2 + (v_2^*)^2 + \sqrt{3}v_1^* v_2^*} \tag{2.32}$$

Thus, if we choose $A \geq (1/2)\sqrt{(v_1^*)^2 + (v_2^*)^2 + \sqrt{3}v_1^* v_2^*}$, statement is proved.

The existence conditions of a sliding mode resulting above is used below for sliding mode design in dynamic systems with control redundancy.

2.3 SLIDING MODE DESIGN

Based on the sufficient existence of the sliding mode condition for systems with redundant control, which resulted in the above mentioned theorem 2, it is possible to offer a two-step procedure of sliding mode design in such systems (Ryvkin, 1984).

On the first step, by introduction of a fictitious control $v(t)$ of dimension m connected by linear transformation (2.27) with initial control vector $u(t)$, a transition to system with a constant matrix before control is carried out.

In this case, the problem of sliding motion design is solved by classical methods, when the dimension of the controlled variables vector, the dimension of the control vector and the number of discontinuous controls are identical, i.e. the matrix D is nonsingular. The design result is allocation in the new control space $v(t) \in V$ an area of the admissible controls V^* providing sliding mode. This area consists of 2^m subareas $V_p^*, p = \overline{1, 2^m}$ (2.22), each of them defined by a combination of signs of control vector components $v_j(t)$ (2.22) and inequalities on their module values (2.21).

These subareas, on the second step, are projected on initial control space U with use of linear transformation (2.18). This control space contains a set of realized control vectors $\tilde{U} = \{U_r\}, r = \overline{1, 2^q}$ received using transformation (2.4) of discontinuous controls $\tilde{u}_i(t)$ (2.3). It is obvious that the modules of discontinuous controls $\tilde{u}_i(t)$ and their switching control must be selected so that at any time each of the above mentioned projected subareas has at least one of the realized control vectors. If the subarea has some vectors, it is necessary to formulate an additional condition of a choice.

For a case considered in consequence from theorem 2, when the structure of a periodic matrix B and control redundancy is defined, sliding mode will exist, if the module of discontinuous control gets out according to (2.32), and the switching control looks like this:

$$sgn \, S_i = sgn(k \, sgn \, Z_2 \cos \gamma_i - sgn \, Z_1 \sin \gamma_i) \tag{2.33}$$

where k is a sign factor,

$$\sqrt{\frac{32u_0^2 + 9V_1^2 + 3\sqrt{3}V_1\sqrt{64u_0^2 - 9V_1^2}}{116u_0^2 - 9V_1^2 - 3\sqrt{3}V_1\sqrt{64u_0^2 - 9V_1^2}}}$$

$$\leq k \leq \sqrt{\frac{116u_0^2 - 9V_2^2 - 3\sqrt{3}V_2\sqrt{64u_0^2 - 9V_2^2}}{32u_0^2 + 9V_2^2 + 3\sqrt{3}V_2\sqrt{64U_0^2 - 9V_2^2}}} \tag{2.34}$$

where $\gamma_1 = at$, $\gamma_2 = at - 2\pi/3$, $\gamma_3 = at + 2\pi/3$.

According to the presented control, operating signals $\operatorname{sgn} S_i$ are connected with fictitious operating signal $\operatorname{sgn} Z_1$ and $\operatorname{sgn} Z_2$ by equation (2.33). Hence, there is a possibility to assign ranges of a angle value γ_i in which the sign of an operating control $\operatorname{sgn} S_i$ coincides with the sign of a fictitious operating control $\operatorname{sgn} Z_1$ or $\operatorname{sgn} Z_2$ and does not depend on the sign of the second control. The border angle values γ_i, at which the control sign is changed, are defined by the formula:

$$\gamma_i = \phi^* \frac{\operatorname{sgn} Z_2}{\operatorname{sgn} Z_1} + (1 + l)\pi \tag{2.35}$$

where $l = 0; \pm 1; \pm 2; \ldots, \phi^* = \arctan(k)$.

Intervals of angle γ_i, on which the sign control $\operatorname{sgn} S_i$ is defined by a fictitious sign control $\operatorname{sgn} Z_1$ or $\operatorname{sgn} Z_2$, depend on values of a factor k. The factor k accordingly (2.34) depends on the values of modules of fictitious controls V_1, V_2 and the module of discontinuous control u_0 selected by the control design. The last gets out, proceeding from an inequality (2.32) received from existence conditions of sliding mode on crossing of surfaces $Z_1 = 0$ and $Z_2 = 0$.

It follows from (2.34), when $u_0 > u_0^*$, there is ambiguity in the choice of values for k and, consequently, the angle γ_i value, according to (2.35), in which a change of sign control takes place. At the minimum permissible value of the modulus of a discontinuous control, i.e. at $u_0 = u_0^*$, the interval of admissible values of the degenerates in the value

$$k^* = \frac{\sqrt{3}\chi + 1}{\chi + \sqrt{3}} \tag{2.36}$$

where $\chi = V_1/V_2$.

With this factor value, there is a unique choice of angle γ_i defining change of the control sign, $\operatorname{sgn} S_i$.

Note that the range of variation of χ between 0 to ∞ at various values of the fictitious controls V_1, V_2 selected in the first stage, corresponds a range of angles for φ^* from $\pi/6$ to $\pi/3$.

In connection with the foregoing, it is useful in sliding mode design to set the sign factor k equal to k^*. In this case it is possible to use the value of the control vector modules that is the minimum allowable one under this method of organization of the sliding motion. The sign control at any given time is assigned one of four possible fictitious sign controls, $\operatorname{sgn} Z_1$, $\operatorname{sgn} Z_2$, $-\operatorname{sgn} Z_1$, $-\operatorname{sgn} Z_2$.

It should be noted that the designed sliding mode control is robust and does not impose high requirements to the constancy of the modulus of the control vector and to determine the angular position γ_1 of admissible control area V^*.

The sliding mode will remain even if the module value of discontinuous control u_0 changes largely. The unique restriction according to inequality (2.32) is the bottom limit of its change u_0^*. As regards the estimation of angular position, it should be defined to be within one of 12 sectors. According to (2.33) and (2.34) by $u_0 > u_0^*$ the accuracy of delimitation between sectors do not show high requirements.

The proposed approach is the key to solving the problem of discontinuous control design for widespread three-phase synchronous electrical drive, being representatives of the specified class of nonlinear systems. Three phase electrical drives have many advantages, namely, their energy transfer efficiency, their simplicity of obtaining a high quality control with the rotating magnetic field, a low impact when changing the parameters of the control plant, the discontinuous control module and external disturbances, the simplicity to measure their angular rotor position, etc.

Furthermore, this approach will be specified in relation to the control problem of a synchronous electrical drive based on a complex "semiconductor power converter – synchronous motor", specific to the different types of semiconductor power converter and synchronous motor.

REFERENCES

Edwards, C. and Spurgeon, S.R. "*Sliding mode control: theory and applications*". London: Taylor & Francis, 1998. 237 p.

Emelyanov, S., Utkin, V., Taran, V., Kostyleva, N., Shubladze, A., Eserov, V. and Dubrovski, E. "*Theory of variable structure control systems*". Moscow: Nauka, 1970. 592 p. (in Russian).

Ryvkin, S. "*Sliding mode in the dynamic system of the special type*". Determinate and stochastic system. Moscow: Nauka, 1984. pp. 38–44. (in Russian).

Utkin, V. "*Sliding modes and their application in variable structure systems*". Moscow: Mir Publ., 1978. 368 p.

Utkin, V. "*Sliding modes in control and optimization*". Berlin: Springer-Verlag, 1992. 286 p.

Utkin, V., Shi J. and Gulder J. "*Sliding modes in electromechanical systems*". London: Taylor & Francis, 1999. 344 p.

Zinober, A.S. "*Variable structure and Lyapunov control*". Berlin: Springer-Verlag, 1994. 420 p.

Chapter 3

State vector estimation

3.1 INFORMATION ASPECTS OF SLIDING MODE DESIGN

The problem of controlling a dynamic plant is inextricably linked with a problem of obtaining estimations of some parts of a state vector. In most cases there is no possibility of directly measuring all components of the state vector that are needed for the control design. A possible way to solve this problem is based on obtaining estimates of unmeasured components of the state vector by constructing special dynamical systems known as observers. This method generates the need for additional analysis of the behavior of extended dynamical system. The additional dynamics can lead to the occurrence of movements in the enhanced system, which differ from those of the original system if there is accurate information on all components of the state vector.

Problems of information support of dynamic system control in the light of use of sliding modes will be considered below. It is possible to define two groups of control problems related to sliding motion. The first one is receiving the necessary state variable information, necessary for realization of sliding movement in a control loop. The second one is using sliding modes to receive the necessary state variable information.

3.2 USE OF AN ASYMPTOTICAL OBSERVER OF THE STATE VARIABLES

It is well known that sliding modes are very sensitive to information misrepresentation of the state vector components received using differentiators. The reason is that sliding motion contains a high frequency component and small differentiator time constants leading to an essential decrease in the frequency of the control switching, which in turn leads to self-oscillations and a decrease in the system accuracy (Emelyanov et al, 1970), (Utkin, 1978), (Utkin, 1992), (Utkin et al, 1999).

A possible way of acquiring information about the derivatives of the state vector components required for the design of a sliding mode, is the use of asymptotical observers of the state vector. They estimate and restore its components based on the measured observable variables (Consoli, 2000), (Dote, 1988), (Krstic et al, 1995), (Kwakernaak & Sivan, 1972), (Luenberger, 1966). The asymptotical observer allows providing an ideal sliding mode, when a dynamic discrepancy between a real dynamic plant and its model takes place. Thus, all attractive properties in a dynamic system, inherent in sliding mode systems, exist even when there is only partial information about the state vector components (Bondarev et al, 1985) due to the measurements conditions.

Let us consider a linear dynamic system of the following kind:

$$\frac{dx(t)}{dt} = Ax + Bu \tag{3.1}$$

where $x(t)$ is the state vector, $x(t) \in R^n$; $u(t)$ is the control vector, $u(t) \in R^m$; A is a matrix of dimension $n \times n$; B is a matrix of dimension $n \times m$, $rank\ B = m$. Matrixes $\{A, B\}$ are controllable.

For the solution of the control problem there is a discontinuous control vector $\breve{u}$, $\breve{u}(t) \in R^m, i = 1, m$. Each control component is defined as:

$$u_i(x, t) = \begin{cases} u_i^+(x, t) & if\ S_i(x, t) > 0 \\ u_i^-(x, t) & if\ S_i(x, t) < 0 \end{cases} \tag{3.2}$$

It is switched on its switching surface $S_i(x) = 0$, which is selected from a condition of the control problem solving $Z = 0$. The switching function vector $S^T = (S_1, \ldots, S_m)$ is a linear function of the state vector x:

$$S = Cx \tag{3.3}$$

where C is a matrix of dimensions $m \times n$. It is related to an error vector of operated variables by the non-singular transformation:

$$S = \tilde{C}Z \tag{3.4}$$

where $\tilde{C}$ is a matrix of dimensions $m \times m$.

A feature of the given problem is that there is information only about a vector of output variables $y(t)$, $y(t) \in R^f$, and $f < n$.

$$y = Fx \tag{3.5}$$

where F is a matrix of dimension $f \times n$, $rank\ F = f$; $\{A, F\}$ are observable matrixes.

For state vector estimation, the linear asymptotical observer is

$$\frac{d\hat{x}(t)}{dt} = A\hat{x} + B\breve{u} + L(F\hat{x} - y) \tag{3.6}$$

where $\hat{x}(t)$ is a state vector of the observer $\hat{x}(t) \in R^n$; L is a constant observer gain matrix of dimension $n \times f$.

Besides, small non-modeled dynamics in the system are present in the control plant:

$$\mu \frac{dq(t)}{dt} = Qq + B_u \breve{u} \tag{3.7}$$

where $q(t)$ is a vector of non considered state variables, $q \in R^r$; Q is a Hurwitz matrix of dimensions $r \times r$; B_u is a matrix of dimensions $r \times m$, $rank\ B_u = r$; μ is the singular parameter defining this non considered dynamics motion. In the measurement channel of the components of a output variable vector $y(t)$ there are small non-modeled dynamics too

$$\eta \frac{dh(t)}{dt} = Hh + B_x x \tag{3.8}$$

where $h(t)$ is an output variable vector of the gauge, $h \in R^e$; H is a Hurwitz matrix of dimension $e \times e$; B_x is a matrix of dimension $e \times n$, $rank\ B_x = e$; η is a singular parameter defining the measuring device dynamics.

As a result the control vector $u(t)$ contains both discontinuous and continuous components

$$u(t) = Pq + R\breve{u} \tag{3.9}$$

where P is a matrix of dimension $m \times r$, $rank\ P = r$; R is a matrix of dimension $m \times m$, $rank\ R = m - r$; $-PQ^{-1}B_u + R = E$, and E is an identity matrix.

Finally a variable vector output has the following form

$$y(t) = P_y h + R_y x \tag{3.10}$$

where P_y is a matrix of dimension $f \times e$, $rank\ P_y = e$; R_y is a matrix of dimension $f \times n$, $rank\ R_y = f - e$; $F = -P_y H^{-1}B_x + R_y$.

Theorem: In the dynamic system defined by (3.1), (3.2), (3.5), (3.6), (3.8), (3.10) there is a multidimensional sliding movement on the crossing of switching surfaces $S = 0$, if it exists in the reduced dynamic system (3.1), (3.2), without being considered small dynamics and having the information of all state vector x components.

Proof: For the solution on the existence problem of a sliding mode we will consider a projection of investigated dynamic system (3.1), (3.2), (3.5), (3.6), (3.8), (3.10) on the switching functions space:

$$S = C\hat{x} \tag{3.11}$$

which is drawn from the estimates of the state vector components, obtained by means of an asymptotic observer:

$$\frac{dS}{dt} = CA\hat{x} + CB\breve{u} + CL(F\hat{x} - P_y h - R_y \hat{x} + R_y \varepsilon) \tag{3.12}$$

where ε is a error vector between a state vector and its estimation $\varepsilon = \hat{x} - x$.

The existence of sliding mode in the reduced dynamic system (3.1), (3.2) unequivocally guarantees its existence and also in the extended investigated one, because matrix

CB as vector control is nonsingular. The difference will consist in the selected value of discontinuous control components, due to occurrence of additional terms of the form $CL(F\hat{x} - P_yh + R_y\hat{x} - R_y\varepsilon)$ in equation (3.12).

To describe the behavior of the extended dynamic system in sliding mode, a method of equivalent control was used. The extended dynamic system was rewritten taking into account the equivalent control:

$$u_{eq} = -(CB)^{-1}C[(A - LP_yH^{-1}B_x)x + (A + LF)\varepsilon - LP_yh] \tag{3.13}$$

$$\frac{dx(t)}{dt} = [E - BR(CB)^{-1}C]A + BR(CB)^{-1}CLP_yH^{-1}B_x]x$$
$$- BR(CB)^{-1}C(A + LF)\varepsilon + BPq + BR(CB)^{-1}CLP_yh \tag{3.14}$$

$$\frac{d\varepsilon(t)}{dt} = \{BPQ^{-1}B_u(CB)^{-1}C[(A - LP_yH^{-1}B_x) - LP_yH^{-1}B_x\}x$$
$$+ [(E + BPQ^{-1}B_u(CB)^{-1}C)A + BPQ^{-1}B_u(CB)^{-1}CLF + LF]\varepsilon$$
$$- BPq - LP_yh \tag{3.15}$$

$$\mu\frac{dq(t)}{dt} = -B_u(CB)^{-1}C[(A - LP_yH^{-1}B_x)x + B_u(A + LF)\varepsilon B + Qq - B_uLP_yh \tag{3.16}$$

$$\eta\frac{dh(t)}{dt} = B_xx + Hh \tag{3.17}$$

While there is sliding mode on switching surfaces, i.e. $S = C\hat{x} = 0$, there are additional algebraic equations that are combined with the state vector components. Thanks to this m state vector, components can be excluded from the system description and the total dynamics system order is reduced to $2n + f + e - m$

$$\frac{dx_1(t)}{dt} = A_1x + B_1\varepsilon + Q_1q + H_1h \tag{3.18}$$

$$\frac{d\varepsilon(t)}{dt} = \{BPQ^{-1}B_u(CB)^{-1}C[(A - LP_yH^{-1}B_x) - LP_yH^{-1}B_x\}x$$
$$+ [(E + BPQ^{-1}B_u(CB)^{-1}C)A + BPQ^{-1}B_u(CB)^{-1}CLF + LF]\varepsilon$$
$$- BPq - LP_yh \tag{3.19}$$

$$\mu\frac{dq(t)}{dt} = -B_u(CB)^{-1}C[(A - LP_yH^{-1}B_x)x + B_u(A + LF)\varepsilon B + Qq - B_uLP_yh \tag{3.20}$$

$$\eta\frac{dh(t)}{dt} = B_xx + Hh \tag{3.21}$$

where x_1 is the state vector describing the movement of the dynamic system (3.1) in sliding mode and A_1, B_1, Q_1, H_1 are the matrixes describing the movement of the dynamic system (3.1) in sliding mode.

Character of movement in a sliding mode is defined by the eigenvalues of the dynamic system (3.18)–(3.21). They can be appointed with use of methods of modal control by selecting suitable factors of matrixes C and L. Thus it is necessary to say that matrixes Q and H are Hurwitz matrixes.

The task is considerably simplified if we take into account the fact that parameters μ and η are small. In this case, the methods of the theory of singularly perturbed equations (Krstic et al, 1995), (Tikhonov, 1952), (Kokotovic, 1976) allow us to decompose the task of analysis and design of motion in the dynamical system (3.18)–(3.21) into a combination of two tasks that differ in the rate of movement. The fast dynamic system is described by equations (3.20) (3.21), and the basic slow system, is described by equations (3.18) (3.19). The dynamics of each of the systems is defined by eigenvalues.

Thus, the use of an asymptotical observer of a state vector is a methodological basis of control design in sliding modes. Its use allows not only to obtain the derivatives and state vector components needed for the design of sliding mode control, but also to compensate for the inaccuracies of the plant model in the high frequency area. This is achieved by closing the feedback through the observer, and it is mainly due to its high-frequency component, which gives structural properties.

3.3 NONLINEAR SLIDING MODE OBSERVER

As it was specified above, information acquisition on an operating process and on components of a state vector in particular, is a central problem for the design of high quality dynamic systems.

The classical approach based on direct measurement of all necessary coordinates, now practically comes to naught. This is caused due to the complexity of direct information acquisition on some components of the state vector, and the general complication of plant that inevitably leads to its operational deterioration and increase in cost indexes. A possible way of overcoming these demerits is to eliminate the sensors of such coordinates (since direct measurement is undesirable), and instead, replace their readings by estimations, obtained by state vector observers.

The complexity of this approach is due to the fact that the majority of the plants are nonlinear and to obtain measurements of its input and output variables, estimates of the state vector components are necessary to develop new nonlinear methods for estimating a component of the state vector. The solution to this problem requires using methods on the theory of systems with sliding modes (Utkin, 1978), (Utkin, 1992), (Utkin et al, 1999), which allow the use of dynamic models with discontinuous parameters, instead of continuously adjusted dynamic models, used for estimation. It allows fulfilling effective breakdown of an observation problem at the expense of the organization of multidimensional sliding movement. It is possible, in this case, to receive an estimation of the unknown state vector components on values a component of equivalent controls of dynamic model. The given approach leads to essential simplification of circuit realization that is rather essential; especially if it would be taking into account plant nonlinearities and its high order (Ryvkin, 2007).

Let us consider a nonlinear dynamic system of the following kind:

$$\frac{dx_1(t)}{dt} = f_1(x_1, t) + D(x_1, t)x_2(t) + B(x_1, t)u(t) \tag{3.22}$$

$$\frac{dx_2(t)}{dt} = f_2(x, u, t) \tag{3.23}$$

where $x(t)^T = (x_1^T, x_2^T)$ is a state vector, $x(t) \in R^n$, $x_1(t) \in R^q$; $f_1(x_1, t)$, $f_2(x, u, t)$ are control plant vectors-columns, $f_1(x_1, t) \in R^q$, $f_2(x, u, t) \in R^{(n-q)}$; $u(t) \in R^m$ is a control; $B(x, t)$ is a matrix of dimension $q \times m$; $D(t)$ is a matrix of dimension $q \times (n - q)$.

State vector x_2 enters linearly into the first dynamic system described by the equation (3.22) and it can be considered a vector of parameters. It is possible, in this case, for this state vector estimation to use the approach, based on a method of "fast" identification on sliding modes (Broslavskii & Schubladze, 1980).

Theorem: The state vector x_2 of nonlinear dynamic system (3.22), (3.23) can be estimated, if the dimension of the measured state vector x_1 is equal or greater to the dimension of an estimated state vector x_2, $q \geq (n - q)$ and in a matrix $D(x_1, t)$ the square matrix $\hat{D}(x_1, t)$, rank $\hat{D}(x_1, t) = (n - q)$ can be selected by all $x_1(t)$.

Proof: If conditions are fulfilled, it is possible to select a subsystem of the following type in the initial dynamic system (3.22):

$$\frac{dx_m(t)}{dt} = f_m(x_1, t) + \hat{D}(x_1, t)x_2(t) + B_m(x_1, t)u(t) \tag{3.24}$$

where $x_m(t)$ is a state vector of a subsystem, $x_m(t) \in R^{n-q}$, $x_1(t)^T = (x_m, x_m^*)$; $f_m(x_1, t)$ is a control plant vector-column of plant, $f_m(x_1, t) \in R^{n-q}$, $f_1(x_1, t)^T = (f_m(x_1, t), f_m^*(x_1, t))$; $B_m(x_1, t)$ is a matrix of dimension $(n - q) \times m$, $B^T(x_1, t) = (B_m(x_1, t), B_m^*(x_1, t))$.

Let us construct a dynamic model of this type:

$$\frac{d\hat{x}_m(t)}{dt} = f_m(x_1, t) + \hat{D}^*(t)v(t) + B_m(x_1, t)u(t) \tag{3.25}$$

where $\hat{x}_m(t)$ is a state vector of model, $\hat{x}_m(t) \in R^{n-q}$; $\hat{D}^*(t)$ is a square matrix with rank $\hat{D}^*(t) = (n - q)$; $v(t)$ is a vector of discontinuous controls of the model, $v(t) \in R^{n-q}$, and its components $v_i(t)$, $i = \overline{1, (n - q)}$, switch on its switching surface $S_i(x, \hat{x}) = 0$ in an expanded space of real and modeled system states.

$$v_i(x, t) = \begin{cases} v_i^+(x, t) & \text{if } S_i(x, t) > 0 \\ v_i^-(x, t) & \text{if } S_i(x, t) < 0 \end{cases} \tag{3.26}$$

Let us require that the model behavior is the same as the subsystem (3.24) behavior, i.e. changes of model state vector $\hat{x}_m(t)$ are the same as ones of subsystem state vector $x_m(t)$. One possible way to control it is to organize it is with use of the discontinuous control vector $v(t)$, the multidimensional sliding movement on crossing the sliding surfaces representing a zero error between the components of model state vector and that of the subsystem.

$$S(x_m, \hat{x}_m) = \hat{x}_m - x_m = \varepsilon = 0 \tag{3.27}$$

In order to prove that the intersection of (3.27) from any random initial conditions is reachable and the movement on this crossing is sustainable, we use the possibility of application of the second method of stability of Lyapunov (Utkin, 1978), (Utkin, 1992), (Utkin et al, 1999). If the derivative of continuously differentiable positive function is negative everywhere except inside the switching surfaces, where it is not defined, then there will be a reaching from any initial conditions at the intersections of the switching surfaces, and sliding motion on this intersection will take place.

Let us choose a positive definite Lyapunov function of this kind

$$W = \frac{1}{2}\varepsilon^T\varepsilon \tag{3.28}$$

Then a derivative of Lyapunov function owing to the equations (3.24), (3.25) is

$$\frac{dW}{dt} = \sum_{i=1}^{n-q} \varepsilon_i \frac{d\varepsilon_i}{dt} \tag{3.29}$$

It follows from (3.29) that the existence condition of multidimensional sliding mode will be fulfilled, if the sum made of component-specific products of error between the components of model state vector and of the system and its derivative, are negative everywhere, except in the crossing of the switching surfaces. The derivative of an error vector ε owing to the equation of movement of initial system (3.24) and dynamic model (3.25) could be written as follows

$$\frac{d\varepsilon}{dt} = -\hat{D}(x_1,t)x_2(t) + \hat{D}^*v(t) \tag{3.30}$$

According to the problem statement, the matrix $\hat{D}^*(t)$ has a full rank. For simplification of the problem solving, it is possible to take it as a unitary matrix. In this case, the sign of each component specific product of the ith error vector components ε_i and its derivative will be always negative, if the algorithm of discontinuous control v_i change is

$$v_i = -V_i \operatorname{sgn} S_i \tag{3.31}$$

and the module of discontinuous control V_i is selected according to the following condition

$$V_i \geq \left| \sum_{j=1}^{n-q} d_{ij}x_{2j} \right| \tag{3.32}$$

where d_{ij} is an element of matrix $\hat{D}(x_1,t)$; x_{2j} is a component of the state vector x_2.

Thus, if conditions (3.31) and (3.32) for all components of the model vector control $v(t)$ are fulfilled, there will be a multi-dimensional sliding motion on the intersection of discontinuous switching surfaces $S(x_m, \hat{x}_m) = \hat{x}_m - x_m = \varepsilon = 0$. The same sliding motion is described using the equivalent control v_{eq}, which is calculated from the equation $dS/dt = 0$ and represents a continuous analogue of discontinuous control

$v(t)$, i.e., its average in sliding mode. In this case, the equation $dS/dt = 0$ looks like:

$$\hat{D}(x_1, t)x_2(t) - v_{eq}(t) = 0 \tag{3.33}$$

Vector-matrix equation (3.33) represents a system $(n - q)$ of linear equations with $(n - q)$ unknown variables. Since on the matrix formulation problem, $\hat{D}(x_1, t)$ is non-singular at all $x_1(t)$, the system has a unique solution for vector x_2:

$$x_2(t) = (\hat{D}(x_1, t))^{-1}v_{eq}(t) \tag{3.34}$$

Thus, the statement is proved, since by using the known value of the equivalent control v_{eq}, the measured values of a state vector x_1 and the known elements of a square matrix $\hat{D}(x_1, t)$ according to (3.34) the value of a state vector x_2 can be calculated.

A nonlinear observer for the state vector x_2, designed using the above mentioned approach, requires additional calculations to obtain an estimate of the state vector components using the received value of the equivalent control vector veq. It is possible to eliminate these calculation by selecting the appropriate type of matrix $\hat{D}^*(t)$ and switching functions. In this case, the equality of the components of the equivalent control vector and those of the state vector could be ensured. The theorem consequences presented below this design follows.

Consequence 1. The state vector x_2 of the nonlinear dynamic system (3.22), (3.23) can be directly estimated by use of equivalent control, if conditions of the theorem are satisfied and the matrix $\hat{D}^*(t)$ is chosen as a matrix $\hat{D}(x_1, t)$.

Proof: If the matrix $\hat{D}^*(t)$ looks like $\hat{D}(x_1, t)$, then according to (3.30) the error ε equation is in sliding mode

$$\frac{d\varepsilon}{dt} = \hat{D}(x_1, t)(-x_2(t) + v(t)) = 0 \tag{3.35}$$

and the equivalent control vector v_{eq} is component specific equal to the state vector $x_2(t)$:

$$v_{eq}(t) = x_2(t) \tag{3.36}$$

Let us prove that from any initial conditions, the system (3.24), (3.25) will get on crossings (3.27) and will steadily move on this crossing. We will use a new vector of error functions $S^{*T} = (S_1^*, \ldots, S_{n-q}^*)$ to prove it, which is related by the linear transformation to the initial error function S

$$S^* = R_S S \tag{3.37}$$

where $R_S = (\hat{D}(x_1, t))^T$ is a matrix of linear transformation.
Owing to the linearity of the transformation $S = S^* = 0$.

In this case the derivative of the chosen Lyapunov function (3.29) taking into account (3.35) can be written down as

$$\frac{dW}{dt} = \varepsilon^T \frac{d\varepsilon}{dt} = (R_S^{-1} S^*)^T \hat{D}(x_1, t)(-x_2(t) + v(t)) \tag{3.38}$$

From this, it follows that

$$\frac{dW}{dt} = (S^*)^T(-x_2(t) + v(t)) \tag{3.39}$$

If the module of discontinuous control V_i to choose

$$V_i \geq |v_{i\,eq}| = |x_2^i| \tag{3.40}$$

where x_2^i is the i-th component of a state vector $x_2(t)$, and the algorithm of control looks like

$$v_i = -V_i \, \text{sgn} \, S_i^* \tag{3.41}$$

that the existence condition of multidimensional sliding movement (negativity of derivatives of the Lyapunov function everywhere, except crossing of switching of ruptures) will be executed. I.e. at any initial conditions, the sliding mode on crossing of switching surfaces $S = S^* = 0$ will take place.

Sliding movement, as it was above mentioned, is described with use of equivalent control v_{eq}, which is calculated from the equation $d\varepsilon/dt = 0$ (3.35) and represents continuous analog of discontinuous control $v(t)$, i.e. its averaged value in a sliding mode. In this case the equation $d\varepsilon/dt = 0$ looks like:

$$x_2(t) - v_{eq}(t) = 0 \tag{3.42}$$

The vector equation (3.42) represents a system of the $(n - q)$ linear equations with $(n - q)$ unknown variables. In the process, each of them contains only one unknown variable, i.e. the components value of the equivalent control vector v_{eq} is component-specific equal to components value of state one $x_2(t)$:

$$v_{eq}(t) = x_2(t) \tag{3.43}$$

Thus, the statement is proved, since the components values of equivalent control vector v_{eq} directly give the components value of a state vector x_2.

Consequence 2. The state vector x_2 of the nonlinear dynamic system (3.22), (3.23) can be directly estimated by using the equivalent control without application of linear transformation if the theorem conditions are satisfied, the matrix $\hat{D}^*(t)$ is chosen as matrix $\hat{D}(x_1, t)$ and a new error vector $S^{*T} = (S_1^*, \ldots, S_{n-q}^*)$ is formed by the additional dynamic system:

$$\frac{dS^*}{dt} = -S^* + R_S u(t) \tag{3.44}$$

where $u^T(t) = (u_1, \ldots, u_{n-q})$, $u_i = \text{sgn}(\varepsilon_i)$, $i = \overline{1, (n - q)}$.

Proof: It is required to prove that in system of an order $2(n - q)$ (3.35), (3.40), (3.41), (3.44) by any initial conditions there is a sliding mode on crossing of switching surfaces $S = S^* = 0$.

Let us consider the positive definite Lyapunov function

$$W = \sum_{i=1}^{n-q} |\varepsilon_i| + \sum_{j=1}^{n-q} |V_j S_j^*| + \sum_{i=1}^{n-q} x_2^j S_j^* \tag{3.45}$$

Outside the switching surfaces the derivative of the Lyapunov function is defined and written according to equations (3.35), (3.40), (3.41), (3.44)

$$\frac{dW}{dt} = \left(\frac{d\varepsilon}{dt}\right)^T \text{sgn } \varepsilon + V^T \frac{dS^*}{dt} + x_2^T \frac{dS^*}{dt} = -(V + x_2)^T S^* \tag{3.46}$$

where $V^T = (V_1 \text{ sgn } S_1^*, \ldots, V_{n-q} \text{ sgn } S_{n-q}^*)$, $(\text{sgn } \varepsilon)^T = (\text{sgn } \varepsilon_1, \ldots, \text{sgn } \varepsilon_{n-q})$.

Equations (3.45) and (3.46) imply that the Lyapunov function derivative is negative defined outside the switching surfaces, due to (3.40). If the trajectory does not cross the switching surfaces or if they are crossed in separate points of a zero measure the Lyapunov function W will decrease monotonically according to this equation $dW/dt + W = 0$.

In case of sliding motion on surfaces intersection, i.e. the $S = S^* = 0$ condition is satisfied only for a part of surfaces

$$S^p = 0 \tag{3.47}$$

$$S^{*r} = 0 \tag{3.48}$$

where $1 \le p \le (n - q)$ and $1 \le r \le (n - q)$ are the number of surfaces on which there is a sliding mode. System movement is described with use of a equivalent control method (2.10), (2.11). The order of the expanded system of the equations (3.35), (3.46) goes down at the expense of an exception a component of error vectors ε and S^* owing to algebraic equations (3.47) and (3.48). Instead of components of discontinuous control vectors $v(t)$ and $u(t)$, their equivalent values were received because equations $dS^p/dt = 0$, $dS^{*r}/dt = 0$ were used.

In this case the variables concerning whether the sliding mode exists can be excluded from the Lyapunov function (3.35). Thus, the function is again differentiated and its derivative is calculated according to the reduced equations system (3.35), (3.46)

$$\frac{dW}{dt} = \sum_{\substack{i=1 \\ i \neq p}}^{n-q} \left(\frac{d\varepsilon_i}{dt} \text{ sgn } \varepsilon_i\right) + \sum_{\substack{j=1 \\ j \neq r}}^{n-q} (V_j \text{ sgn } S_j^* + x_2^j) = -(V + x_2)^T S^* \tag{3.49}$$

where $V^T = (V^1, V^2)$ is a vector with the components

$$(V^1)^T = (V_1 \text{ sgn } S_1^*, \ldots, V_{n-q-p} \text{ sgn } S_{n-q-p}^*),$$

$$(V^2)^T = (v_{(n-q-p)eq}, \ldots, v_{(n-q)eq}), S^* \in R^{n-q-r}$$

Obviously, as in this case the Lyapunov function derivative is negative out of the switching surfaces and monotonously decreases, and the expanded system (3.35), (3.46) tends asymptotically to multidimensional sliding movement on crossing of surfaces $S = S^* = 0$.

3.4 PHYSICAL SIGNIFICANCE OF EQUIVALENT CONTROL

As it is known (Utkin, 1978), (Utkin, 1992), (Utkin et al, 1999), the equivalent control is defined through limiting transition from a real sliding mode in $\Delta-$ in the vicinities of switching surfaces crossings $S = 0$ to the ideal one:

$$\lim_{\substack{\tau \to 0 \\ \Delta/\tau \to 0}} u_{av} = u_{eq} \tag{3.50}$$

where the u_{av} is the average value of discontinuous control $v(t)$ received with the use of the filter

$$\tau \frac{du_{av}}{dt} + u_{av} = v(t) \tag{3.51}$$

with a time constant τ.

Thus, when constructing the observer it is necessary to consider this feature, i.e. the presence of a filter with its own dynamics. Therefore, when analyzing the requirements imposed on the expanded observable nonlinear system (3.22), (3.23), (3.51) and (3.25), it is necessary to consider that the estimation of a component of a state vector x_2 is possible in case that the filter is singular perturbed in relation to subsystem (3.22), i.e. the rates of a filtration should be essentially above (by an order of magnitude) the rates of change of the estimated components of a state vector.

The proposed information approaches are a key to solving the design problem of the discontinuous controls for widespread three-phase electrical drive, being representatives of the specified class of nonlinear systems. Presented approaches to the information support problem solving open wide possibilities for designing modern electrical drives with high operational reliability and control quality, thanks to sensors number reduction, control simplification and its robustness to variations of plant parameters, of the value of the source voltage and external drive loads and to the measurements assurance of rotor angular position. Furthermore, this approach will be specified taking into account specificity of various types of synchronous motors.

REFERENCES

Bondarev, A.G., Bondarev, S.A., Kostyleva, N.E. and Utkin, V.I. *"Sliding modes in systems with asymptotic state observers"*. Automation and Remote Control, 1985, vol. 46, no. 6, pp. 679–684.

Broslavskii, A.D. and Shubladze, A.M. *"Solution to problem of 'fast' identification using multidimensional sliding modes"*. Automation and Remote Control, 1980, vol. 41, no. 2, part 1, pp. 195–199.

Consoli, A. *"Advanced control techniques. Modern Electrical Drives"*. Dordrecht, Boston, London: Kluwer Academic Publishers, 2000, pp. 523–582.

Dote, Y. *"Application of modern control techniques to motor control"*. Proc. of the IEEE, 1988, vol. 76, no. 4, pp. 438–445.

Emelyanov, S., Utkin, V., Taran, V., Kostyleva, N., Shubladze, A., Eserov, V. and Dubrovski, E. *"Theory of variable structure control systems"*. Moscow: Nauka, 1970. 592 p. (in Russian).

Kokotovic, P.V., O'Malley, R.B. and Sannuti, P. *"Singular perturbation and reduction in control theory"*. Automatica, 1976, no. 12, pp. 123–132.

Krstic, M., Kanellakopoulos, I. and Kokotovic, P. *"Nonlinear and Adaptive Control Design"*. New York: Wiley, 1995. 563 p.

Kwakernaak, H. and Sivan, R. *"Linear optimal control systems"*. New York: John Wiley & Son Inc., 1972. 608 p.

Luenberger, D.C. *"Observers for multivariable systems"*. IEEE Transactions on Automatic Control, 1966, vol. 11, no. 1, pp. 190–197.

Ryvkin, S. *"Estimation of state vector components in a nonlinear singular system"*. Control Sciences, 2007, no. 4, pp. 8–13. (in Russian).

Tikhonov, N. *"Systems of differential equations with a small parameter multiplying derivations"*. Mathematicheskii Sbornik, 1952, vol. 73, no. 31, pp. 575–586. (in Russian).

Utkin, V. *"Sliding modes and their application in variable structure systems"*. Moscow: Mir Publ., 1978. 368 p.

Utkin V. *"Sliding modes in control and optimization"*. Berlin: Springer-Verlag, 1992. 286 p.

Utkin, V., Shi, J. and Gulder, J. *"Sliding modes in electromechanical systems"*. London: Taylor & Francis, 1999. 344 p.

Chapter 4

Synchronous drive control design

4.1 SINGLE-LOOP CONTROL DESIGN

4.1.1 The two step decomposition approach

As it was explained in section 1.2, the basic controlled variable in an electrical drive is usually the angular speed $\Omega(t)$, which ideally should be equal to its reference $\Omega_z(t)$. In a synchronous motor, used in electrical drives, there are two independent operating influences outside the stator. Creating the basic magnetic flux by an excitation winding from outside the rotor opens up an additional possibility to control the electrical drive along with angular speed control. It is possible to achieve better electrical drive performance, e.g., by getting the largest possible efficiency while keeping the power factor equal to one. The desirable electrical drive mode can be achieved by selecting the appropriate reference for the control variables.

Using such references, i.e. a mechanical $\Omega_z(t)$ or electrical $\omega_z(t)$ rotor angular speed one and the stator current $i_{dz}(t)$, it is possible to achieve specific functioning requirements of the synchronous motor. Also, it is handy to control the excitation current $i_{fz}(t)$ if there is an excitation winding.

Having chosen those references, the electrical drive behavior is described by the error functions of the control variables, which are deviations of the actual values of the control variables from their references:

$$
\begin{aligned}
Z_1 &= C_1(\omega_z - \omega) + \frac{d(\omega_z - \omega)}{dt}, \\
Z_2 &= i_{dz} - i_d, \\
Z_3 &= i_{fz} - i_f
\end{aligned}
\tag{4.1}
$$

where C_1 is a constant coefficient.

Solving the electrical drive control problem means designing controls, which are output voltages of the three-phase semiconductor power converter that feeds the synchronous motor stator windings and the single-phase semiconductor power converter feeding the excitation winding, makes the above error function (4.1) equal to zero.

In this case, currents i_d and i_f are equal to their reference values i_{dz} and i_{fz} respectively, and the error between the reference and the actual rotor angular speed by

$C_1 > 0$ tends exponentially to zero according to a law having a time constant equal to $1/C_1$. One of the possible variants of maintenance of simultaneous equality to zero of functions (4.1) is the organization of a sliding mode at surfaces crossings $Z_1 = 0$, $Z_2 = 0$ and $Z_3 = 0$. A direct solution for this problem, among known methods, encounters a problem of control redundancy. In this connection the two-step design procedure, which is based on a sufficient condition of existence of sliding movement in such systems, presented in chapter two, has been developed using decomposition. The initial problem of control design is divided into tasks of smaller complexity that are solved separately. A suggested two-step approach takes working features of the synchronous motor and semiconductor power converter separately into account. At the first step, only features of synchronous motors of various types are considered. A design problem of a sliding motion is solved by using a fictitious control vector and a classical mathematical model of the synchronous motor (see Park's equations presented in Section 1.1.1). The fictitious controls are two-phase voltages of the biphasic Park model. They do not really exist, however such model is very convenient for control design. Unlike them, the real motor control is a three phase voltage formed by the phase switching controls of the converter.

The second step carries out the transition to actual controls taking into account functioning features and peculiarities of various types of the three-phase semiconductor power converter.

4.1.2   First step – design of fictitious discontinuous control

As mentioned in chapter two, the question on the possibility of organizing sliding movement and designing the necessary controls is solved using equation (2.5), which describes a motion projection of the initial dynamic system, in our case a synchronous motor, in an error subspace of the controlled variables Z. When using the classical Park's description for the synchronous motor and the control variables specified above, the projection equation is as follows:

$$\frac{dZ}{dt} = F + AU \tag{4.2}$$

where vector $F = (dz_z/dt) - Gf$ and matrix $A = GB$ do not depend on the control vector U, but are defined by the column vector $f(x, t)$ and the control matrix $B(x, t)$ of the synchronous motor, respectively. The existing conditions directly depend on the kind of matrix A.

The design of sliding modes control varies depending on the type of synchronous motor, the structure, and the value of the elements of a control matrix $B(x, t)$ and column-vector $f(x, t)$, and both the structure and the elements of matrix A. We will start with the control design for a salient-pole synchronous motor with an excitation winding (Ryvkin, 1982).

Salient-pole synchronous motor with an excitation winding

The motor is described by the fourth order differential equations (1.1), (1.4) with a fictitious control vector $U^T = (u_q, u_d, u_f)$, whose components u_d, u_q are voltage on the fictitious two-phase stator winding in a rotating coordinate system (d, q) and voltage across the excitation winding u_f. The error function vector $Z^T = (Z_1, Z_2, Z_3)$ (4.1) contains all the three components. Vector F and matrix A inserted into equation (4.2) look like these:

$$
F = \left(
\begin{array}{c}
-\dfrac{1}{J}
\left\{
\begin{array}{l}
\left[(L_d - L_q)\left(C_1 - \dfrac{r}{L_q}\right) - r\left(1 - \dfrac{L_f L_d}{L_1^2}\right) \right] i_d\, i_q \\[2ex]
+ L_{df}\left(C_1 - \dfrac{r}{L_q} - \dfrac{r_f L_q}{L_1^2}\right) i_f i_q \\[2ex]
- \left[\dfrac{1}{L_q}(L_d i_d + L_{df} i_f)(L_d i_q - L_q i_d + L_{df} i_f) - L_q\left(1 - \dfrac{L_f L_q}{L_1^2}\right) i_q^2 \right]\omega \\[2ex]
- \left(C_1 M + \dfrac{dM}{dt}\right) + J\left(C_1 \dfrac{d\omega_z}{dt} + \dfrac{d^2\omega_z}{dt^2}\right)
\end{array}
\right\} \\[6ex]
-\dfrac{1}{L_1^2}(-L_f r i_d + L_{df} r_f i_f + L_f L_q \omega i_q) + \dfrac{di_{dz}}{dt} \\[3ex]
-\dfrac{1}{L_1^2}(-L_d r_f i_f + L_{df} r i_d - L_{df} L_q \omega i_q) + \dfrac{di_{fz}}{dt}
\end{array}
\right)
\tag{4.3}
$$

$$
A = \left(
\begin{array}{ccc}
-\dfrac{L_{df} i_f + (L_d - L_q)i_d}{J L_q} & -\dfrac{L_f(L_d - L_q) - L_{df}^2}{J L_1^2} i_q & -\dfrac{L_{df} L_q}{J L_1^2} i_q \\[3ex]
0 & -\dfrac{L_f}{L_1^2} & \dfrac{L_{df}}{L_1^2} \\[3ex]
0 & \dfrac{L_{df}}{L_1^2} & -\dfrac{L_d}{L_1^2}
\end{array}
\right)
\tag{4.4}
$$

After consideration of matrix A it is obvious that the component u_q of the control vector U does not influence the occurrence of a sliding mode on surfaces $Z_2 = 0$ and $Z_3 = 0$. Hence, it is possible, using a method of control hierarchy (2.9), to carry out decomposition of an initial design problem into two independent problems of smaller dimension, having organized the hierarchy of occurrence of sliding movement on crossing of switching surfaces $Z_1 = 0$, $Z_2 = 0$ and $Z_3 = 0$:

1. Design of two-dimensional sliding mode on crossing surfaces $Z_2 = 0$ and $Z_3 = 0$.
2. Design of one-dimensional sliding mode on surface $Z_1 = 0$.

To solve the first problem it is necessary to consider the fact that the synchronous motor own inductance is always larger that its mutual inductance ($L_f > L_{df}$, $L_d > L_{df}$) (Leonhard, 2001), (Boldea & Nasar, 2005), (Pahman & Zhou, 2000). Taking this into account, the matrix before control in the truncated system is stationary and has a prevailing diagonal. Then, according to a sufficient condition of existence of a sliding mode (2.8) the sliding mode on crossing of surfaces $Z_2 = 0$ and $Z_3 = 0$ will exist in the event that controls u_d and u_f change under the following law:

$$u_d = u_{d0} \operatorname{sgn} Z_2 \tag{4.5}$$

$$u_f = u_{f0} \operatorname{sgn} Z_3 \tag{4.6}$$

The value of their amplitudes will exceed the modules of their corresponding equivalent controls:

$$u_{d0} \geq |u_{deq}| = \left| r i_d - L_q \omega i_q + L_d \frac{di_{dz}}{dt} + L_{df} \frac{di_{fz}}{dt} \right| \tag{4.7}$$

$$u_{f0} \geq |u_{feq}| = \left| r_f i_f + L_{df} \frac{di_{dz}}{dt} + L_f \frac{di_{fz}}{dt} \right| \tag{4.8}$$

In this case, sliding movement on each of the switching surfaces takes place.

In solving the control design problem, provided a sliding motion on the surface $Z_1 = 0$, according to the method of control hierarchy, it is assumed that the sliding mode is already taking place on the intersection of surfaces $Z_2 = 0$ and $Z_3 = 0$. In this case, in the equation describing system sliding mode on the surface $Z_1 = 0$ according to the method of equivalent control, controls u_d and u_f are replaced with their equivalent controls u_{deq} (4.7) and u_{feq} (4.8), obtained by solving the first problem of sliding motion design. According to a existence condition of a one-dimensional sliding motion (2.6), sliding mode on the surface $Z_1 = 0$, and, hence, on the crossing of surfaces $Z_1 = 0$, $Z_2 = 0$ and $Z_3 = 0$ will exist, if control u_q varies according to the following law:

$$u_q = u_{q0} \operatorname{sgn} Z_1 \tag{4.9}$$

and its amplitude exceeds the module of its corresponding equivalent control u_{qeq}

$$u_{q0} \geq |u_{qeq}| = \left| \begin{array}{l} -(C_1 L_q - r)i_q + (L_d i_d + L_{df} i_{fz})\omega + \dfrac{L_q \left(C_1 M + \frac{dM}{dt} \right)}{L_{df} i_{fz} + (L_d - L_q)i_{dz}} \\[4mm] + \dfrac{J L_q \left(C_1 \frac{d\omega_z}{dt} + \frac{d^2\omega_z}{dt^2} \right)}{L_{df} i_{fz} + (L_d - L_q)i_{dz}} - \dfrac{L_q \left(L_{df} \frac{di_{fz}}{dt} + (L_d - L_q) \frac{di_{dz}}{dt} \right) i_q}{L_{df} i_{fz} + (L_d - L_q)i_{dz}} \end{array} \right| \tag{4.10}$$

It is necessary to notice that on "reasonable" operating modes, the magnetic flux $\Psi = L_{df} i_{fz} + (L_d - L_q)i_{dz}$, causing electromagnetic torque M_{el}, has to be different to zero, otherwise there would be electromagnetic torque and no control possibility for the synchronous motor. Hence, at $\Psi \neq 0$ there is a basic possibility of performance of a condition (4.10) at the expense of a choice of value u_{q0} equal or surpassing equivalent control u_{qeq}, i.e. maintenance of existence of a sliding mode along surface $Z_1 = 0$.

The existence conditions of the sliding mode received above are based on a necessary and sufficient condition of existence of the one-dimensional sliding mode, therefore if the received inequalities (4.7), (4.8), (4.10) are carried out for all points of a system state space (4.2), they define not only the conditions of existence of sliding movement on crossing surfaces $Z_1 = 0$, $Z_2 = 0$ and $Z_3 = 0$, but also on the reached ones.

Equations (4.5), (4.6), (4.9) and inequalities (4.7), (4.8), (4.10) describe the control, and the above mentioned inequalities at $Z = 0$ define a class of admissible perturbations and the references reproduced without dynamic tracking errors.

It is necessary to notice that the control of an excitation current (4.6), (4.8), obtained on the first design step, first, provides maintenance of a excitation current at the reference value, and, second, is designed for a real, physically existing, excitation voltage and can be realized with the use of a full-bridge DC-DC converter with symmetrical switching. The excitation winding is included in the bridge diagonal (figure 4.1).

In this case, there is a source of constant voltage u_{f0}. The excitation winding ends can be connected to source line (+) by means of switches K_1 and K_2, depending on the control switches p_{K1} and p_{K2}. Also, they can be connected to source line (−) by means of switches K_3 and K_4, depending on the control switches p_{K3} and p_{K4}. If the control signal is +1, it means that the corresponding switch is closed. Otherwise, if it is −1, the switch is open.

In accordance to the symmetric switching law, the switches located in a diagonal of the pulse converter, i.e. $p_{K1} = p_{K3} = -p_{K2} = -p_{K4} = \text{sgn}\,Z_3$ switch on simultaneously. The designed control (4.6) and the value of the necessary feed voltage u_{f0} defined by inequality (4.8) are set.

The direction and magnitude of the excitation flux in the synchronous motor generated by the excitation current is usually known. In this case it does not make sense to use the power supply circuit presented in figure 4.1, but it is preferable to use the single-phase DC-DC converter, which is one arm of the bridge scheme: switch pair K_1 and K_4 or K_2 and K_3. Depending on the control signal $p_{K1} = -p_{K4} = \text{sgn}\,Z_3$ ($p_{K2} = -p_{K3} = -\text{sgn}\,Z_3$) switches are connected at one end of the excitation winding to line of a constant voltage source u_{f0} (figure 4.2). The second end excitation winding is connected to a zero voltage source.

In this case, the voltage on the excitation winding will represent a time sequence of unipolar impulses, whose duration is defined by switch-on time of switch K_1. Moreover, the necessary voltage to maintain sliding mode is defined by the

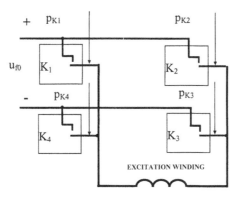

Figure 4.1 Full-bridge DC-DC converter with excitation winding

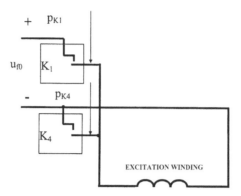

Figure 4.2 Single-phase DC-DC converter with excitation winding

following formula:

$$u_f = \frac{u_{f0}}{2}(1 + \text{sgn } S_3) \tag{4.11}$$

In this case, the power supply circuit of excitation winding shown above, along with ease of implementation, provides the best spectral composition of excitation voltage compared with the bridge scheme.

Nonsalient-pole synchronous motor with excitation winding

From the matrix structure viewpoint describing the motor behavior in the space of switching function Z, nothing differs from the salient-pole synchronous motor with the excitation winding considered above. The only distinction is the matrix elements, since the design procedure and the obtained algorithms are equivalent to the results shown above. The difference is shown depending on selecting conditions of control vector component amplitudes.

Taking into account (1.7), vector F and matrix A look like these:

$$F = \left(\begin{array}{l} -\dfrac{1}{H}\left\{ \begin{array}{l} r\dfrac{L_{df}}{L_1^2}i_d i_q + L_{df}\left(C_1 - \dfrac{r}{L_q} - \dfrac{r_f L}{L_1^2}\right)i_f i_q \\[2mm] -\left[\dfrac{1}{L}(Li_d + L_{df}i_f)(Li_q - Li_d + L_{df}i_f) - L\left(1 - \dfrac{L_f L}{L_1^2}\right)i_q^2\right]\omega \\[2mm] -\left(C_1 M + \dfrac{dM}{dt}\right) + H\left(C_1\dfrac{d\omega_z}{dt} + \dfrac{d^2\omega_z}{dt^2}\right) \end{array} \right\} \\[10mm] -\dfrac{1}{L_1^2}(-L_f r i_d + L_{df}r_f i_f + L_f L\omega i_q) + \dfrac{di_{dz}}{dt} \\[4mm] -\dfrac{1}{L_1^2}(-Lr_f i_f + L_{df} r i_d - L_{df}L\omega i_q) + \dfrac{di_{fz}}{dt} \end{array} \right) \tag{4.12}$$

$$A = \begin{pmatrix} -\dfrac{L_{df}}{HL_q}i_f & \dfrac{L_{df}^2}{HL_1^2}i_q & -\dfrac{L_{df}L}{HL_1^2}i_q \\[2mm] 0 & -\dfrac{L_f}{L_1^2} & \dfrac{L_{df}}{L_1^2} \\[2mm] 0 & \dfrac{L_{df}}{L_1^2} & -\dfrac{L}{L_1^2} \end{pmatrix} \qquad (4.13)$$

The control components are

$$u_q = u_{q0}\,\mathrm{sgn}\,Z_1 \qquad (4.14)$$

$$u_d = u_{d0}\,\mathrm{sgn}\,Z_2 \qquad (4.15)$$

$$u_f = u_{f0}\,\mathrm{sgn}\,Z_3 \qquad (4.16)$$

Inequalities on a choice of amplitudes of control components are

$$u_{q0} \geq |u_{qeq}| = \left| \begin{aligned} &-(C_1 L - r)i_q + (L i_d + L_{df}i_{fz})\omega + \frac{L\left(C_1 M + \frac{dM}{dt}\right)}{L_{df}i_{fz}} \\ &+\frac{HL\left(C_1\frac{d\omega_z}{dt} + \frac{d^2\omega_z}{dt^2}\right)}{L_{df}i_{fz}} - \frac{L_q\frac{di_{fz}}{dt}i_q}{i_{fz}} \end{aligned} \right| \qquad (4.17)$$

$$u_{d0} \geq |u_{deq}| = \left| r i_d - L\omega i_q + L\frac{di_{dz}}{dt} + L_{df}\frac{di_{fz}}{dt} \right| \qquad (4.18)$$

$$u_{f0} \geq |u_{feq}| = \left| r_f i_f + L_{df}\frac{di_{dz}}{dt} + L_f\frac{di_{fz}}{dt} \right| \qquad (4.19)$$

As well as in the previous case, the obtained inequalities (4.17)–(4.19) define not only the sliding movement existence conditions on the surfaces crossings $Z_1 = 0$, $Z_2 = 0$ and $Z_3 = 0$, but also the condition to reach these crossings. Second, these inequalities at $Z = 0$ define a class of admissible perturbations and the references reproduced without a dynamic tracking error. The designed control for an excitation winding can be implemented – as well as in the previous case – using DC-DC converters.

Salient-pole permanent magnet synchronous motor

This motor can be modeled by a differential equation system of third order (1.1), (1.5), where $U^T = (u_q, u_d)$ is a fictitious control vector, whose components u_d, u_q are voltage on a fictitious biphasic winding in a rotating coordinate system (d, q) (Ryvkin, 1991). The error function vector $Z^T = (Z_1, Z_2)$ contains only two components. In this case, vector F and matrix A can be combined into equation (4.2).

$$
F = -\frac{1}{J}\left(
\begin{array}{c}
\left\{
\begin{array}{l}
(L_d - L_q)\left(C_1 - \dfrac{r}{L_d} - \dfrac{r}{L_q}\right)i_d i_q + \left(C_1 - \dfrac{r}{L_q}\right)\Psi_f i_q \\[2mm]
-\left[\dfrac{1}{L_q}(L_d i_d + \Psi_f)(L_d i_q - L_q i_d + \Psi_f)\right]\omega \\[2mm]
+\left[\dfrac{L_q}{L_d}(L_d - L_q)i_q^2\right]\omega - \left(C_1 M + \dfrac{dM}{dt}\right) + J\left(C_1\dfrac{d\omega_z}{dt} + \dfrac{d^2\omega_z}{dt^2}\right)
\end{array}
\right\} \\[10mm]
\dfrac{r}{L_d}i_d - \dfrac{L_q}{L_d}\omega i_q + \dfrac{di_{dz}}{dt}
\end{array}
\right)
$$

$$
\text{(4.20)}
$$

$$
A = \left(
\begin{array}{cc}
-\dfrac{1}{J}\left(\dfrac{L_d - L_q}{L_d}\right)i_q & -\dfrac{1}{J}\left[\left(\dfrac{L_d - L_q}{L_q}\right)i_d + \Psi_f\right] \\[4mm]
-\dfrac{1}{L_d} & 0
\end{array}
\right)
$$

$$
\text{(4.21)}
$$

After matrix A inspection, it is visible that component u_q of a control vector U does not influence sliding mode occurrence on surface $Z_2 = 0$. Hence, it is possible again, using a method of control hierarchy (2.9), to carry out decomposition of an initial design problem on two independent one-dimensional problems on the existence of one-dimensional sliding modes on surfaces $Z_1 = 0$ and $Z_2 = 0$. The control design and a choice of the control amplitude, for each of them, are carried out with the use of existence conditions of one-dimensional sliding movement. The sliding motion is designed first on surface $Z_2 = 0$ using u_d, and then on surface $Z_1 = 0$ using control u_q.

The control

$$
u_d = u_{d0}\,\mathrm{sgn}\,Z_2 \tag{4.22}
$$

and the inequality

$$
u_{d0} \geq |u_{deq}| = \left| r i_d - L_q \omega i_q + L_d \frac{di_{dz}}{dt}\right| \tag{4.23}
$$

provide contrast of signs on error function Z_2 and its derivative, i.e. the sliding mode performance for the above specified existence conditions for a scalar case.

By solving the control design problem for the second component of a control vector u_q, the principle of control hierarchy is used. It is supposed that sliding mode on surface $Z_2 = 0$ takes place, and it is necessary to provide sliding mode on surface $Z_1 = 0$. In this case, according to a method of equivalent control (2.10), in the first equation of system (4.2) instead of control u_d, its value, namely the equivalent control u_{deq}, is substituted. Sliding mode on surface $Z_1 = 0$, and, hence, and on the crossings of surfaces $Z_1 = 0$ and $Z_2 = 0$ will take place, if

$$
u_q = u_{q0}\,\mathrm{sgn}\,Z_1 \tag{4.24}
$$

and

$$u_{q0} \geq |u_{qeq}| = \left| \begin{array}{l} -(C_1 L_q - r)i_q + (L_d i_d + \Psi_f)\omega + \dfrac{L_q\left(C_1 M + \frac{dM}{dt}\right)}{\Psi_f + (L_d - L_q)i_{dz}} \\[4mm] +\dfrac{JL_q\left(C_1\frac{d\omega_z}{dt} + \frac{d^2\omega_z}{dt^2}\right)}{[\Psi_f + (L_d - L_q)i_{dz}]} - \dfrac{L_q(L_d - L_q)i_q}{\Psi_f + (L_d - L_q)i_{dz}}\dfrac{di_{dz}}{dt} \end{array} \right| \qquad (4.25)$$

It is important to notice that on "reasonable" operating modes the magnetic flux $\Psi = \Psi_f + (L_d - L_q)i_{dz}$ causing electromagnetic torque M_{el} has to be different to zero, otherwise there would be no possibility to control the electromagnetic torque of the synchronous motor. Hence, at $\Psi \neq 0$ there is a basic possibility of performance of condition (4.25) at the expense of a choice of a large enough value u_{q0}, i.e. maintenance of sliding mode existence along the surface $Z_1 = 0$.

As a result at the first stage, decomposition of an initial design problem is carried out and two one-dimensional problems about the occurrence of sliding modes are considered independently. The received equations (4.22), (4.24) describe a working control, and an inequality (4.23), (4.25) at $Z = 0$ define a class of admissible perturbations and the references reproduced without a dynamic tracking error.

The *nonsalient-pole* permanent magnet synchronous motor and the synchronous reluctance motor, considered below, are described the same as a salient-pole permanent magnet synchronous motor, with a differential equation system of third order, according to (1.1), (1.8) and (1.1), (1.8) with the same fictitious control vectors $U^T = (u_q, u_d)$ and error functions $Z^T = Z_1, Z_2)$ (Ryvkin, 1995). Matrix A, before control in the equation (4.2), has the same structure, as well as in the case of a salient-pole synchronous motor. The only distinction is the value of matrix elements. Therefore the design procedure and the received control are equivalent to the presented above for a salient-pole permanent magnet synchronous motor. There are differences only in the conditions of a choice of amplitude of the control vector components.

Nonsalient-pole permanent magnet synchronous motor

$$F = \left(\begin{array}{c} -\dfrac{1}{H}\left\{ \begin{array}{l} \left(C_1 - \dfrac{r}{L_q}\right)\Psi_f i_q - \left[\dfrac{1}{L}(Li_d + \Psi_f)(Li_q - Li_d + \Psi_f)\right]\omega \\[3mm] -\left(C_1 M + \dfrac{dM}{dt}\right) + J\left(C_1\dfrac{d\omega_z}{dt} + \dfrac{d^2\omega_z}{dt^2}\right) \end{array} \right\} \\[8mm] \dfrac{r}{L}i_d - \omega i_q + \dfrac{di_{dz}}{dt} \end{array} \right) \qquad (4.26)$$

$$A = \left(\begin{array}{cc} 0 & -\dfrac{\Psi_f}{J} \\[4mm] -\dfrac{1}{L_d} & 0 \end{array} \right) \qquad (4.27)$$

The controls are:

$$u_d = u_{d0} \operatorname{sgn} Z_2 \tag{4.28}$$

$$u_q = u_{q0} \operatorname{sgn} Z_1 \tag{4.29}$$

The selected amplitudes of the control vector components u_d and u_q are:

$$u_{d0} \geq |u_{deq}| = \left| ri_d - L\omega i_q + L\frac{di_{dz}}{dt} \right| \tag{4.30}$$

$$u_{q0} \geq |u_{qeq}| = \left| \begin{array}{l} -(C_1 L_q - r)i_q + (Li_d + \Psi_f)\omega + \dfrac{L\left(C_1 M + \frac{dM}{dt}\right)}{\Psi_f} \\[4mm] + \dfrac{JL_q\left(C_1 \frac{d\omega_z}{dt} + \frac{d^2\omega_z}{dt^2}\right)}{\Psi_f} \end{array} \right| \tag{4.31}$$

Synchronous reluctance motor

$$F = -\frac{1}{J} \left\{ \begin{array}{l} \left[(L_d - L_q)\left(C_1 - \dfrac{r}{L_d} - \dfrac{r}{L_q}\right)i_d i_q + \left(C_1 - \dfrac{r}{L_q}\right)\Psi_f i_q \right] \\[3mm] -\left[\dfrac{L_d}{L_q}(L_d i_q - L_q i_d) \right]i_d\omega + \left[\dfrac{L_q}{L_d}(L_d - L_q)i_q^2 \right]\omega \\[3mm] -\left(C_1 M + \dfrac{dM}{dt} \right) + J\left(C_1 \dfrac{d\omega_z}{dt} + \dfrac{d^2\omega_z}{dt^2} \right) \\[3mm] \dfrac{r}{L_d}i_d - \dfrac{L_q}{L_d}\omega i_q + \dfrac{di_{dz}}{dt} \end{array} \right\} \tag{4.32}$$

$$A = \begin{pmatrix} -\dfrac{1}{J}\left(\dfrac{L_d - L_q}{L_d}\right)i_q & -\dfrac{1}{J}\left[\left(\dfrac{L_d - L_q}{L_q}\right)i_d\right] \\[4mm] -\dfrac{1}{L_d} & 0 \end{pmatrix} \tag{4.33}$$

The controls are:

$$u_d = u_{d0} \operatorname{sgn} Z_2 \tag{4.34}$$

$$u_q = u_{q0} \operatorname{sgn} Z_1 \tag{4.35}$$

The selected amplitudes of the components u_d and u_q are:

$$u_{d0} \geq |u_{deq}| = \left| ri_d - L_q\omega i_q + L_d\frac{di_{dz}}{dt} \right| \tag{4.36}$$

$$u_{q0} \geq |u_{qeq}| = \left| \begin{array}{l} -(C_1 L_q - r)i_q + L_d i_d \omega + \dfrac{L_q \left(C_1 M + \frac{dM}{dt} \right)}{(L_d - L_q)i_{dz}} \\[4mm] +\dfrac{J L_q \left(C_1 \frac{d\omega_z}{dt} + \frac{d^2 \omega_z}{dt^2} \right)}{(L_d - L_q)i_{dz}} - \dfrac{L_q i_q}{i_{dz}} \dfrac{di_{dz}}{dt} \end{array} \right| \qquad (4.37)$$

It is necessary again to notice that on "reasonable" operating modes for magnetic flux $\Psi = (L_d - L_q)i_{dz}$ conditioning electromagnetic torque M_{el} is nonzero, since otherwise there is no possibility of controlling the electromagnetic torque of the synchronous motor. Consequently, when $\Psi \neq 0$ there is a theoretical possibility, by choosing a sufficiently large value of u_{q0} to ensure the fulfillment of conditions for the existence of one-dimensional sliding motion on the surface $Z_1 = 0$.

Equations (4.28), (4.29) and (4.34), (4.35), describe the algorithm of control, and the corresponding pair of inequalities (4.30), (4.31) and (4.36), (4.37) at $Z = 0$ define a class of admissible perturbations and references, reproduced without a dynamic tracking error.

Thus, regardless of the type of synchronous motor, the control that supports control execution, i.e. keeping a zero value of the error function (4.1), contains independent control loops. There are always two control loops outside the stator, which provide an angular speed $\Omega(\omega)$ control and a current stator component i_d control:

$$u_d = u_{d0} \operatorname{sgn} Z_2 \qquad (4.38)$$

$$u_q = u_{q0} \operatorname{sgn} Z_1 \qquad (4.39)$$

There is one more control loop in the presence of an excitation winding

$$u_f = u_{f0} \operatorname{sgn} Z_3 \qquad (4.40)$$

Figure 4.3 presents the control structure. Specific features of various types of synchronous motors find the reflection in a choice of value of discontinuous voltage amplitudes u_{d0}, u_{q0}, u_{f0}. Thus, the results of the first design step are:

- Physically realized control of the excitation winding (4.40);
- Allocation in two-dimensional fictitious control space of admissible control areas U^* providing sliding mode existence.

The above mentioned region of admissible control U^* consist of four sub-regions $U_1^*, U_2^*, U_3^*, U_4^*$ (figure 4.4), allocated in control space according to the designed control (4.38), (4.39) (table 4.1) and inequalities in the choice of amplitude control components u_{d0}, u_{q0}. $U^* = \{U^*\} = U_1^* \cup U_2^* \cup U_3^* \cup U_4^*$, $U_1^* \cap U_2^* = 0$, $U_1^* \cap U_3^* = 0$, $U_1^* \cap U_4^* = 0$, $U_2^* \cap U_3^* = 0$, $U_2^* \cap U_4^* = 0$, $U_3^* \cap U_4^* = 0$.

4.1.3   Second step – phase voltage control design

On the second step, there is a transition from the fictitiously entered control vector $U^T = (u_d, u_q)$ to the real control of a semiconductor power converter feeding the

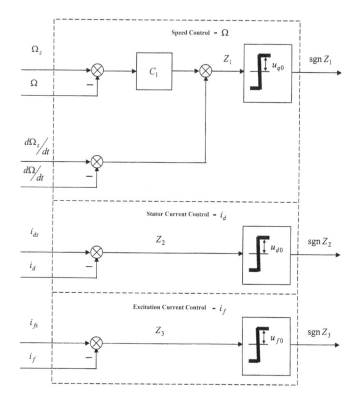

Figure 4.3 Synchronous motor control structure

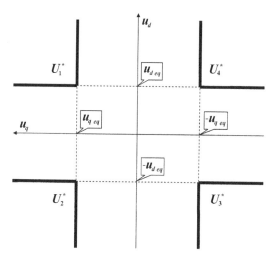

Figure 4.4 Sliding mode control areas $U_1^*, U_2^*, U_3^*, U_4^*$

Table 4.1 Subareas of admissible controls

	U_1^*	U_2^*	U_3^*	U_4^*
sgn Z_1	1	−1	−1	1
sgn Z_2	1	1	−1	−1

synchronous motor. Moreover, our attention is now centered in the semiconductor power converter type, since all types of synchronous motor considered above have the same control, and their difference consists only in the value of the controls amplitudes. Therefore, from now on, the term "synchronous motor" will be used.

As it was emphasized earlier on, in modern synchronous drives, a three phase semiconductor power converter feeds the motor stator windings. It was shown in section 1.1.2 that the semiconductor power converter can accept only 7 output voltage vectors in the case of a voltage source inverter (VSI), or 24 in the case of a matrix converter. Each of these vectors has a fixed direction, and none of them coincides with the formally entered control vector $U^T = (u_d, u_q)$ either in direction or in value. Amplitude values of the components of a formally entered control vector are chosen according to inequalities that are valid for the controls in the rotating coordinate system (d, q).

Equation (1.3) and transformation matrix (1.2) relate controls u_d, u_q, and real phase voltage to obtain the output phase voltages of the semiconductor power converter. In this case, the phase voltages would provide the sliding mode designed at the first step and represent sinusoidal voltages whose frequency is a multiple of the angular speed of the synchronous motor rotor with both variable amplitude and phase. These phase voltage to within a high-frequency component could be obtained on the output of semiconductor power converter by using high-frequency PWM, requiring calculation of the duration of switch-on time and sequence of switching (Holz, 1994), (Kuerker, 2000). However such approach will demand additional calculations connected with the allocation of equivalent controls, necessary for calculation of PWM, and will defeat one of the basic properties of sliding mode, which is the simplicity of implementation.

The alternative approach to design output phase voltages of the semiconductor power converter is based on a sufficient existence condition of a sliding mode in systems with redundant control, exposed in section 2.2. Selecting conditions of amplitude values of formally entered controls are based on inequalities. It allows allocating in space of fictitious controls, in this case u_d, u_q, an area of the admissible controls U^* guaranteeing sliding mode existence. It is obvious that if designing such real control, e.g. three-phase voltage, their total projections to axes (d, q) coincide on a sign with the designed fictitious controls, and the values of projections satisfy sliding mode inequalities, the sliding mode on crossing of chosen surfaces takes place, though the projections value, i.e. the values of the formally entered controls u_d, u_q, will change during the work.

To design such controls, it is necessary to transform the area of admissible controls received in the rotating coordinate system using a well-known linear Park's

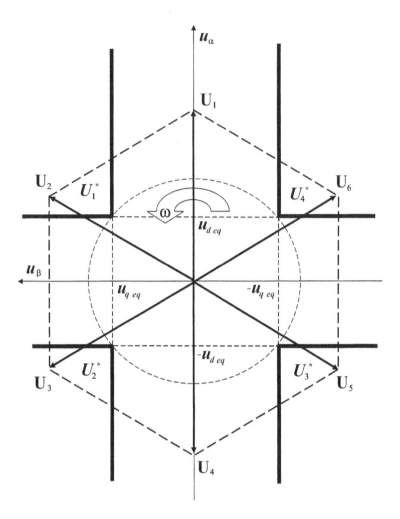

Figure 4.5 Subareas of admissible control U^* and VSI output voltage vectors in a motionless coordinate system u_α, u_β

transformation (Leonhard, 2001), (Boldea & Nasar, 2005).

$$\begin{pmatrix} u_\alpha \\ u_\beta \end{pmatrix} = \left\| \begin{matrix} \cos \gamma_A & -\sin \gamma_A \\ \sin \gamma_A & \cos \gamma_A \end{matrix} \right\| \begin{pmatrix} u_d \\ u_q \end{pmatrix} \tag{4.41}$$

at a fixed coordinate system of the control space (u_α, u_β). Then, in this fixed coordinate system the relative position of this area (4.41) and the output vectors of the semiconductor power converter must be analyzed. The transformed area of admissible controls U^* (figure 4.5) has the same configuration as the original area in the rotating coordinate system (u_d, u_q), but it is not fixed. Instead, it rotates with the electrical rotor speed ω.

It is necessary to mention that transition from two to three operating influences implies ambiguity. In other words, it is possible to offer some equivalent from the control viewpoint of converter controls, at which the existence conditions of a sliding mode are fulfilled. Some possible controls below will be concretely defined, depending on the type of the semiconductor power converter.

Voltage source inverter (VSI). As shown in section 1.1.2 its output voltage vector can accept six nonzero values, separated an angular distance of $\pi/3$ from each other (figure 1.5).

The situation completely coincides with the one considered in section 2.2, which is a special case of solving the problem of a sliding mode existence. There is no doubt, after direct check, that at all the values γ_j ($j = R, S, T$) and any values of sgn Z_1 and sgn Z_2 at each of the subareas of the area of admissible control, have at least one of six nonzero output vectors realized by VSI, if

$$U_{in} \geq U_{in}^* = \frac{3}{2}\sqrt{u_{d0}^2 + u_{q0}^2 + \sqrt{3}u_{d0}u_{q0}} \tag{4.42}$$

$$\text{sgn } S_j = \text{sgn}(k \text{ sgn } S_2 \cos \gamma_j - \text{sgn } S_1 \sin \gamma_j) \tag{4.43}$$

$$\sqrt{\frac{8U_{in}^2 + 9u_{d0}^2 + 3\sqrt{3}u_{d0}\sqrt{16U_{in}^2 - 9u_{d0}^2}}{24U_{in}^2 - 9u_{d0}^2 - 3\sqrt{3}u_{d0}\sqrt{16U_{in}^2 - 9u_{d0}^2}}}$$

$$\leq k \leq \sqrt{\frac{24U_{in}^2 - 9u_{q0}^2 - 3\sqrt{3}u_{q0}\sqrt{16U_{in}^2 - 9u_{q0}^2}}{8U_{in}^2 + 9u_{q0}^2 + 3\sqrt{3}u_{q0}\sqrt{16U_{in}^2 - 9u_{q0}^2}}} \tag{4.44}$$

It is obvious that in this case there is a possibility to allocate ranges of angle values γ_j, at which the sign of the VSI phase switch control p_j coincides with the sign or the opposite one of the fictitious control sgn Z_1 or sgn Z_2 and does not depend on a sign of the second one. Angle values of γ_j, at which there is a sign change of the operating signal p_j, are defined by the formula (2.35).

As it appears from (4.44), there is an ambiguity in the selection of the factor k value by $U_{in} > U_{in}^*$ and, hence, the values of γ_j, according to (2.35), where there is a change of a control sign. At the minimum admissible value of input voltage VSI, i.e. by $U_{in} = U_{in}^*$, the value of factor k is

$$k^* = \frac{\sqrt{3}\chi + 1}{\chi + \sqrt{3}} \tag{4.45}$$

where $\chi = u_{d0}/u_{q0}$, and a single-valued choice of a angle γ_j defining the control p_j rate of change, takes place.

It must be emphasized that a range of angle φ^* changes from $\pi/6$ to $\pi/3$ corresponding to the range of change of χ, from 0 to ∞, by the various values u_{d0}, u_{q0} chosen at the first step.

In connection with the above mentioned, it is expedient to use in sliding mode design a value of sign factor k equal to k^*. In this case the VSI input voltage value is the

Table 4.2 Control signals of VSI phase switches

γ_R	p_R	p_S	p_T
$\left(-\dfrac{\pi}{3}+\varphi^*; \dfrac{\pi}{3}-\varphi^*\right)$	$\dfrac{1+\operatorname{sgn}Z_2}{2}$	$\dfrac{1+\operatorname{sgn}Z_1}{2}$	$\dfrac{1-\operatorname{sgn}Z_1}{2}$
$\left(\dfrac{\pi}{3}-\varphi^*; \varphi^*\right)$	$\dfrac{1+\operatorname{sgn}Z_2}{2}$	$\dfrac{1+\operatorname{sgn}Z_1}{2}$	$\dfrac{1-\operatorname{sgn}Z_2}{2}$
$\left(\varphi^*; \dfrac{2\pi}{3}-\varphi^*\right)$	$\dfrac{1-\operatorname{sgn}Z_1}{2}$	$\dfrac{1+\operatorname{sgn}Z_1}{2}$	$\dfrac{1-\operatorname{sgn}Z_2}{2}$
$\left(\dfrac{2\pi}{3}-\varphi^*; \dfrac{\pi}{3}+\varphi^*\right)$	$\dfrac{1-\operatorname{sgn}Z_1}{2}$	$\dfrac{1+\operatorname{sgn}Z_2}{2}$	$\dfrac{1-\operatorname{sgn}Z_2}{2}$
$\left(\dfrac{\pi}{3}+\varphi^*; \pi-\varphi^*\right)$	$\dfrac{1-\operatorname{sgn}Z_1}{2}$	$\dfrac{1+\operatorname{sgn}Z_2}{2}$	$\dfrac{1+\operatorname{sgn}Z_1}{2}$
$\left(\pi-\varphi^*; \dfrac{2\pi}{3}+\varphi^*\right)$	$\dfrac{1-\operatorname{sgn}Z_2}{2}$	$\dfrac{1+\operatorname{sgn}Z_2}{2}$	$\dfrac{1+\operatorname{sgn}Z_1}{2}$
$\left(\dfrac{2\pi}{3}+\varphi^*; \dfrac{4\pi}{3}-\varphi^*\right)$	$\dfrac{1-\operatorname{sgn}Z_2}{2}$	$\dfrac{1-\operatorname{sgn}Z_1}{2}$	$\dfrac{1+\operatorname{sgn}Z_1}{2}$
$\left(\dfrac{4\pi}{3}-\varphi^*; \pi+\varphi^*\right)$	$\dfrac{1-\operatorname{sgn}Z_2}{2}$	$\dfrac{1-\operatorname{sgn}Z_1}{2}$	$\dfrac{1+\operatorname{sgn}Z_2}{2}$
$\left(\pi+\varphi^*; \dfrac{5\pi}{3}-\varphi^*\right)$	$\dfrac{1+\operatorname{sgn}Z_1}{2}$	$\dfrac{1-\operatorname{sgn}Z_1}{2}$	$\dfrac{1+\operatorname{sgn}Z_2}{2}$
$\left(\dfrac{5\pi}{3}-\varphi^*; \dfrac{4\pi}{3}+\varphi^*\right)$	$\dfrac{1+\operatorname{sgn}Z_1}{2}$	$\dfrac{1-\operatorname{sgn}Z_2}{2}$	$\dfrac{1+\operatorname{sgn}Z_2}{2}$
$\left(\dfrac{4\pi}{3}+\varphi^*; 2\pi-\varphi^*\right)$	$\dfrac{1+\operatorname{sgn}Z_1}{2}$	$\dfrac{1-\operatorname{sgn}Z_2}{2}$	$\dfrac{1-\operatorname{sgn}Z_1}{2}$
$\left(2\pi-\varphi^*; \dfrac{5\pi}{3}+\varphi^*\right)$	$\dfrac{1+\operatorname{sgn}Z_2}{2}$	$\dfrac{1-\operatorname{sgn}Z_2}{2}$	$\dfrac{1-\operatorname{sgn}Z_1}{2}$

minimum possible one at the supposed sliding mode design approach $U_{in} = U_{in}^*$. Moreover, the law of formation of switch controls p_j, depending on γ_R, can be presented in the form of a logic table (table 4.2).

According to the table, for the switch control p_j, one of four signs functions ($\operatorname{sgn}Z_1, \operatorname{sgn}Z_2, -\operatorname{sgn}Z_1, -\operatorname{sgn}Z_2$) is used at any time. The received sliding mode existence conditions, as it was specified above, are sufficient, as at any moment the values u_{d0}, u_{q0} undertake from the sets caused by inequalities on a choice of control amplitudes, i.e. sliding movement on each of surfaces $Z_1 = 0$ and $Z_2 = 0$ takes place. It is necessary to notice that the offered sliding motion control is robust and does not have high requirements besides constancy of VSI input voltage U_{in} and definition of rotor angular position γ_R. The sliding mode will remain even if the value of VSI input voltage U_{in} changes largely. The unique limitation according to inequality (4.42) is the bottom bound of its change U_{in}^*. So it is possible to simplify the scheme of a direct current link and to refuse the smoothing capacity. As to defining angular position, it should be defined all on all to within one of 12 sectors.

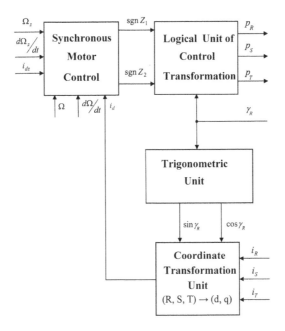

Figure 4.6 VSI drive control

According to (4.43) and (4.44) by $U_{in} > U_{in}^*$ high demands also are not made to border definition between sectors on accuracy.

The possible variant of structural realization of the controller, which ensures the equivalents of the rotor angular speed Ω and its reference $\Omega_z(t)$, and of the stator current component and its reference $i_d = i_{dz}(t)$, is shown in figure 4.6.

Matrix Converter (MC). As shown in section 1.1.2, as opposed to VSI, a matrix converter can generate, during each time, one of the 24 nonzero vectors of output voltage, which are the tops of three hexagons and two three-phase systems (figure 1.7).

One of the possible ways to use the matrix converter to ensure a sliding mode control of a synchronous motor is the use of the output voltage vectors, which are the vertices of a hexagon, i.e. transforming a matrix converter control problem into a voltage source inverter one. From the viewpoint of the maximum utilization of three-phase power supply it is preferable to use the output vectors $U_1, \ldots, U_6$ forming the first hexagon. Moreover, taking into account that the existence conditions of sliding mode with respect to the power converter (4.42)–(4.44) are also based on inequalities and, as it was mentioned above, to ensure sliding motion condition, it is possible to use the pulsing voltage U_{in} bounded only from below by the value of U_{in}^*, for the organization of sliding movement it is possible to use supply line three phase voltage directly. The only restriction is that the smallest value of the module of the output vector of the matrix converter, forming this hexagon, must exceed the value of U_{in}^* obtained by the condition (4.43), i.e. the amplitude of phase voltage is related to U_{in}^*, according to the following equation:

$$U \geq \frac{2}{3} U_{in}^* \tag{4.46}$$

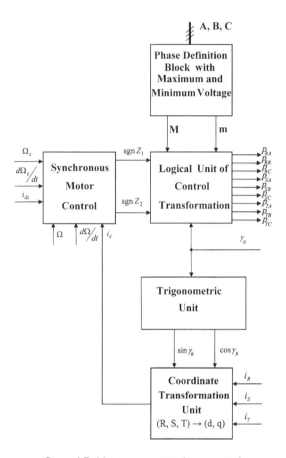

Figure 4.7 Matrix converter drive control

The formation law of switch control p_{ij}, depending on γ_R, in this case includes the definition of phases with minimum and maximum voltage values from three-phase supply and definition of a switch control of a load phase p_i. The last one can be defined according to a logic table (table 4.2). Thus it is necessary to consider that the tabular value $p_i = 0$ corresponds to a switch control m, and the tabular value $p_i = 1$ corresponds to M.

As in the previous case, the sliding mode existence conditions are sufficient at any time. Values u_{d0}, u_{q0} are taken from the sets caused by inequalities, depending on the type of synchronous motor, i.e. sliding movement on each of surfaces $Z_1 = 0$ and $Z_2 = 0$ takes place.

The proposed control for the synchronous drive based on a complex "matrix converter – synchronous motor" is also robust and does not impose high requirements, both to the three-phase supply line voltage and to determine the angular position of the rotor. Sliding mode will remain, even if the amplitude of a three-phase voltage supply line voltage changes largely. The only limitation, according to inequality (4.48), is the lower limit of its change U_{in}^*.

As to the definition of angular position, as well as in the previous case, is to be defined within one of 12 sectors, and according to (4.44) and (4.45), when $U_{in} > U_{in}^*$, high demands on accuracy of the border definition between sectors are not made.

In comparison with the previous case, the control is supplemented with the block of phase definition with minimum and maximum voltage. The possible variant of structural realization of the control providing the equivalents of the rotor angular speed Ω and its reference $\Omega_z(t)$ and of the stator current component and its reference $i_d = i_{dz}(t)$ is shown in figure 4.7.

4.2 CASCADE (SUBORDINATED) CONTROL

As specified in section 1.2, the synchronous electrical drive control can be decomposed in two parts, according to its mechanical and its electrical process dynamics. In this case, the purpose of the electrical drive control is, as before, to maintain the demanded output mechanical characteristics. The semiconductor power converter is in the inner closed loop. Its control must be designed, so that the above mentioned demands will be fulfilled. A slower angular speed controller forms the electromagnetic torque M_{el} reference, which the fast inner electromagnetic loop included "semiconductor power converter – synchronous motor" (figure 4.8) fulfills. Since the value of the electromagnetic torque M_{el} depends on the value of the stator current, the actual inner loop is the current control loop.

Now, as it was specified in the introduction, this is the most widespread control solution, based on the use of multidimensional sliding motion. Unfortunately, such approach to control a complex "voltage source inverter – synchronous motor", has not found the non-ambiguity until today, in the description and the name. For the description of such control, various approaches, sometimes empirical, are used that have found reflection and in variety of the terms used for the name of this control.

The theoretical substantiation of this control type resulted in this section, based using the theory of nonlinear systems with discontinuous controls. The results of this theory allow explaining not only the widely known high quality results obtaining using this voltage source inverter control, but also those problems and complexities, which arise at its realization.

The mathematical model used for the analysis of processes in a complex "semiconductor power converter – synchronous motor", includes only a description of the synchronous motor electromagnetic part, which, as shown in section 1.1.1, depends on the type of synchronous motor.

As well as in the previous case, there is a redundancy control for the sliding mode design, and it is useful to take the advantage of the approach offered in section 2.3 and realized in section 4.1, i.e. the two-step decomposition approach to sliding mode design.

A qualitative difference consists in the choice as one of control variables a component of the stator current $i_q(t)$. In this case the behavior of the complex "semiconductor power converter – synchronous motor" will be characterized by the following error functions of the control variables:

$$Z_1 = i_{qz} - i_q, \quad Z_2 = i_{dz} - i_d, \quad Z_3 = i_{fz} - i_f \qquad (4.47)$$

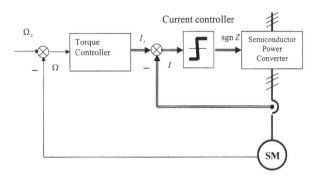

Figure 4.8 Synchronous motor cascade control structure

As specified before, nullification of the control variables errors (4.47) in a sliding mode on surfaces $Z_1 = 0$, $Z_2 = 0$ and $Z_3 = 0$ crossing simultaneously should be provided by switch control design. Switches of a three-phase semiconductor power converter commute output phase voltages, which feed the synchronous motor stator windings. The single-phase semiconductor power converter feeds the excitation winding.

In this case all three currents i_d, i_q and i_f are equal to the references i_{dz}, i_{qz} and i_{fz} respectively.

As it was specified in section 2.3, the problem of the possibility of sliding mode organization and the design of the necessary controls deal with the use of equation (2.5) describing a projection of movement of the initial dynamic system, in our case the synchronous motor at a subspace Z of control variables errors. The classical Park's description of synchronous motor and the specified above control variables are used. The projection equation looks like (4.2).

Since depending on the synchronous motor type and the selected control variables vector $z(t)$, the structure and a element value of a matrix before control $A(x, t)$ and column-vector $F(x, t)$ defining in compliance with a sliding mode existence condition the control and the needed control amplitude are changed. Hence, the control design problem should solve for each type synchronous motor separately. The basic date and designed sliding mode control, both for all types of synchronous motors, presented in section 1.1.1, will be quoted below.

Salient-pole synchronous motor with an excitation winding

Only third order differential equations describing the electromagnetic processes in the synchronous motor are used for control design (1.1). The acting fictitious control vector $U^T = (u_q, u_d, u_f)$ components are the voltage on fictitious two-phase stator winding in rotating coordinate system (d, q) u_d, u_q and voltage across the excitation winding u_f. The error function vector $Z^T = (Z_1, Z_2, Z_3)$ (4.47) contains all three components. In that case, vector F and matrix A in equation (4.2) have the following form:

$$F = \begin{vmatrix} -\dfrac{1}{L_q}(-ri_q - L_{df}\omega i_f - L_d\omega i_d) + \dfrac{di_{qz}}{dt} \\[3mm] -\dfrac{1}{L_1^2}(-L_f ri_d + L_{df}r_f i_f + L_f L_q \omega i_q) + \dfrac{di_{dz}}{dt} \\[3mm] -\dfrac{1}{L_1^2}(-L_d r_f i_f + L_{df} ri_d - L_{df}L_q \omega i_q) + \dfrac{di_{fz}}{dt} \end{vmatrix} \qquad (4.48)$$

$$A = \begin{pmatrix} \dfrac{1}{L_q} & 0 & 0 \\[3mm] 0 & -\dfrac{L_f}{L_1^2} & \dfrac{L_{df}}{L_1^2} \\[3mm] 0 & \dfrac{L_{df}}{L_1^2} & -\dfrac{L_d}{L_1^2} \end{pmatrix} \qquad (4.49)$$

It is obvious from consideration of matrix A (4.49) that the component u_q of a control vector U is responsible for the existence of a sliding mode on surfaces $Z_1 = 0$, and components of a control vector u_d and u_f are responsible for sliding motion at the surfaces crossing $Z_2 = 0$ and $Z_3 = 0$. Hence, the initial design problem of sliding mode on surfaces $Z_1 = 0$, $Z_2 = 0$ and $Z_3 = 0$ crossings breaks up into two independent problems of smaller dimensions:

– Design of one-dimensional sliding mode on surface $Z_1 = 0$.
– Design of two-dimensional sliding mode on surfaces $Z_2 = 0$ and $Z_3 = 0$.

The first design problem solving of control u_q providing sliding mode on surface $Z_1 = 0$ is based on a "classical" existence condition of one-dimensional sliding mode (2.6). Sliding mode on surface $Z_1 = 0$ will exist, if control u_q changes under the following law

$$u_q = u_{q0} \operatorname{sgn} Z_1 \qquad (4.50)$$

and its amplitude will exceed the equivalent control $u_{q\,eq}$

$$u_{q0} \geq |u_{q\,eq}| = \left| ri_q + L_{df}\omega i_f + L_d \omega i_d + L_q \frac{di_{qz}}{dt} \right| \qquad (4.51)$$

To solve the second design problem, it is necessary to consider that the synchronous motor own inductance is always larger that its mutual inductance ($L_f > L_{df}$, $L_d > L_{df}$) (Leonhard, 2001), (Boldea & Nasar, 2005), (Pahman & Zhou, 2000). Taking this into account, the matrix before control in the truncated system is stationary with a prevailing diagonal. Then, according to a sufficient existence condition of a sliding mode (2.8) the sliding mode on crossing of surfaces $Z_2 = 0$

and $Z_3 = 0$ will exist in the event that controls u_d and u_f will change under the following law

$$u_d = u_{d0} \operatorname{sgn} Z_2 \tag{4.52}$$

$$u_f = u_{f0} \operatorname{sgn} Z_3 \tag{4.53}$$

The value of their amplitudes will exceed modules of corresponding equivalent controls

$$u_{d0} \geq |u_{d\,eq}| = \left| r i_d - L_q \omega i_q + L_d \frac{d i_{dz}}{dt} + L_{df} \frac{d i_{fz}}{dt} \right| \tag{4.54}$$

$$u_{f0} \geq |u_{f\,eq}| = \left| r_f i_f + L_{df} \frac{d i_{dz}}{dt} + L_f \frac{d i_{fz}}{dt} \right| \tag{4.55}$$

If above mentioned conditions have been fulfilled the sliding mode has following features:

- Sliding mode takes place on the crossing of switching surfaces $Z_1 = 0$, $Z_2 = 0$ and $Z_3 = 0$.
- The received existence conditions of a sliding mode are based on a necessary and sufficient existence condition of a one-dimensional sliding mode. Hence the resulting inequalities (4.51), (4.54), (4.55). If they hold for all points of state space of the system (4.2), they define the terms not only for the existence of sliding movement on the intersection of surfaces $Z_1 = 0$, $Z_2 = 0$ and $Z_3 = 0$, but also reaching it.
- The relations (4.50), (4.52), (4.53) describe the control.
- Inequalities (4.51), (4.54), (4.55) at $Z_1 = 0$ determine the class of admissible perturbations and the impacts reproduced without the dynamic error tracking.

It should be noticed that the designed excitation current control (4.53), (4.56) as in section 4.1.2, is designed for real physically existing excitation voltage. Its implementation is not different from the control implementation (4.6), (4.8).

All remarks concerning the properties of the designed controls and the sliding mode, stated above, are valid for all the various types of synchronous motor described below.

Nonsalient-pole synchronous motor with excitation winding

From the viewpoint of structure of matrix A and vector F describing motor behavior in space of control variable error or that of the switching function there is the same error vector Z. The mathematical model of this motor do not differ from considered above one of salient-pole synchronous motor with excitation winding. There are distinctions only in matrix elements, therefore the design procedure and the received control will be equivalent to the resulting ones above. The specified difference finds the expression in selection conditions of amplitude of control vector components.

Taking into account (1.7), vector F and matrix A look like this:

$$F = \begin{vmatrix} -\dfrac{1}{L}(-ri_q - L_{df}\omega i_f - L\omega i_d) + \dfrac{di_{qz}}{dt} \\[3mm] -\dfrac{1}{L_1^2}(-L_f ri_d + L_{df} r_f i_f + L_f L\omega i_q) + \dfrac{di_{dz}}{dt} \\[3mm] -\dfrac{1}{L_1^2}(-Lr_f i_f + L_{df} ri_d - L_{df} L_q \omega i_q) + \dfrac{di_{fz}}{dt} \end{vmatrix} \qquad (4.56)$$

$$A = \begin{pmatrix} \dfrac{1}{L} & 0 & 0 \\[3mm] 0 & -\dfrac{L_f}{L_1^2} & \dfrac{L_{df}}{L_1^2} \\[3mm] 0 & \dfrac{L_{df}}{L_1^2} & -\dfrac{L}{L_1^2} \end{pmatrix} \qquad (4.57)$$

From consideration of matrix A (4.57) it is obvious that, as well as in the previous case, the initial problem of sliding mode design on crossing of surfaces $Z_1 = 0$, $Z_2 = 0$ and $Z_3 = 0$ breaks up into two independent problems of smaller dimension.

The controls

$$u_q = u_{q0} \operatorname{sgn} Z_1 \qquad (4.58)$$

$$u_d = u_{d0} \operatorname{sgn} Z_2 \qquad (4.59)$$

$$u_f = u_{f0} \operatorname{sgn} Z_3 \qquad (4.60)$$

and the following inequalities, on a choice of control amplitudes,

$$u_{q0} \ge |u_{q\,eq}| = \left| ri_q + L_{df}\omega i_f + L\omega i_d + L\frac{di_{qz}}{dt} \right| \qquad (4.61)$$

$$u_{d0} \ge |u_{deq}| = \left| ri_d - L\omega i_q + L\frac{di_{dz}}{dt} + L_{df}\frac{di_{fz}}{dt} \right| \qquad (4.62)$$

$$u_{f0} \ge |u_{f\,eq}| = \left| r_f i_f + L_{df}\frac{di_{dz}}{dt} + L_f\frac{di_{fz}}{dt} \right| \qquad (4.63)$$

ensure existence of sliding movement on crossing of surfaces $Z_1 = 0$, $Z_2 = 0$ and $Z_3 = 0$ and their reaching.

Salient-pole permanent magnet synchronous motor

The behavior of this electric motor is described by a second order system of differential equation (1.5). The fictitious control vector $U^T = (u_q, u_d)$ components are fictitious biphasic voltages in rotating coordinate system (d, q) u_d, u_q. The control variable error vector $Z^T = (Z_1, Z_2)$ contains only two components. Vector F and matrix A can be combined into equation (4.2) producing the following expressions

$$F = \begin{vmatrix} -\dfrac{1}{L_q}(-ri_q - \Psi_f\omega - L_d\omega i_d) + \dfrac{di_{qz}}{dt} \\ -\dfrac{1}{L_d}(-ri_d + L_q\omega i_q) + \dfrac{di_{dz}}{dt} \end{vmatrix} \tag{4.64}$$

$$A = \begin{pmatrix} \dfrac{1}{L_q} & 0 \\ 0 & \dfrac{1}{L_d} \end{pmatrix} \tag{4.65}$$

It is obvious from consideration of matrix A (4.65) that component u_q of a control vector U is responsible for existence of a sliding mode on surface $Z_1 = 0$ and a component u_d of a control vector – for existence of sliding movement on surface $Z_2 = 0$. Hence, the initial design problem of sliding mode on crossing of surfaces $Z_1 = 0$ and $Z_2 = 0$ breaks up into two independent one-dimensional problems. Each of them dares with use existence conditions of one-dimensional sliding movement (2.6). In this case, the control

$$u_q = u_{q0}\,\mathrm{sgn}(Z_1) \tag{4.66}$$

$$u_d = u_{d0}\,\mathrm{sgn}(Z_2) \tag{4.67}$$

and inequalities on a selection of control components amplitudes

$$u_{q0} \geq |u_{qeq}| = \left| -ri_q - \Psi_f\omega - L_d\omega i_d + L_q\frac{di_{qz}}{dt} \right| \tag{4.68}$$

$$u_{d0} \geq |u_{deq}| = \left| ri_d - L_q\omega i_q + L_d\frac{di_{dz}}{dt} \right| \tag{4.69}$$

provide existence of sliding movement on crossing of surfaces $Z_1 = 0$ and $Z_2 = 0$ and their reaching.

The electromagnetic parts considered below the synchronous reluctance motor and *nonsalient-pole* permanent magnet synchronous motor are described as the salient-pole permanent magnet one, i.e. with a differential equations system of second order, according to (1.6) and (1.7). Also they have the same vector of fictitious control $U^T = (u_q, u_d)$ and functions of errors $Z^T = (Z_1, Z_2)$, matrix A of the same structure. Therefore procedure of control design is similar resulted above for salient-pole permanent magnet synchronous motor. The existing differences are only the equivalent controls values, which are defined by the value of the elements of matrix A and of column-vector F. For these types of synchronous motors, expressions for matrix A and for column-vector F, and also control and a condition on a choice of amplitude of the controls providing sliding movement on surfaces crossing $Z_1 = 0$ and $Z_2 = 0$ will be shown below.

Nonsalient-pole permanent magnet synchronous motor

$$F = \begin{vmatrix} -\dfrac{1}{L}(-ri_q - \Psi_f\omega - L\omega i_d) + \dfrac{di_{qz}}{dt} \\[2mm] -\dfrac{1}{L}(-ri_d + L\omega i_q) + \dfrac{di_{dz}}{dt} \end{vmatrix} \tag{4.70}$$

$$A = \dfrac{1}{L}\begin{pmatrix} 1 & 0 \\ 0 & 1 \end{pmatrix} \tag{4.71}$$

$$u_q = u_{q0}\,\mathrm{sgn}(Z_1) \tag{4.72}$$

$$u_d = u_{d0}\,\mathrm{sgn}(Z_2) \tag{4.73}$$

$$u_{q0} \geq |u_{qeq}| = \left| -ri_q - \Psi_f\omega - L\omega i_d + L\dfrac{di_{qz}}{dt} \right| \tag{4.74}$$

$$u_{d0} \geq |u_{deq}| = \left| ri_d - L\omega i_q + L\dfrac{di_{dz}}{dt} \right| \tag{4.75}$$

Synchronous reluctance motor

$$F = \begin{vmatrix} -\dfrac{1}{L_q}(-ri_q - L_d\omega i_d) + \dfrac{di_{qz}}{dt} \\[2mm] -\dfrac{1}{L_d}(-ri_d + L_q\omega i_q) + \dfrac{di_{dz}}{dt} \end{vmatrix} \tag{4.76}$$

$$A = \begin{pmatrix} \dfrac{1}{L_q} & 0 \\[3mm] 0 & \dfrac{1}{L_d} \end{pmatrix} \tag{4.77}$$

$$u_q = u_{q0}\,\mathrm{sgn}(Z_1) \tag{4.78}$$

$$u_d = u_{d0}\,\mathrm{sgn}(Z_2) \tag{4.79}$$

$$u_{q0} \geq |u_{q\,eq}| = \left| -ri_q - L\omega i_d + L\dfrac{di_{qz}}{dt} \right| \tag{4.80}$$

$$u_{d0} \geq |u_{d\,eq}| = \left| ri_d - L\omega i_q + L\dfrac{di_{dz}}{dt} \right| \tag{4.81}$$

Thus, without dependence from synchronous motor type the control providing performance of a control problem in view, i.e. maintenance in a sliding mode of zero function of control variables errors (4.47), contains independent control loops of control variables. There are always two control loops of stator current components i_d and i_q in the control

$$u_d = u_{d0}\,\mathrm{sgn}(Z_2) \tag{4.82}$$

$$u_q = u_{q0}\,\mathrm{sgn}(Z_1) \tag{4.83}$$

In the presence of an excitation winding there is one more control loop

$$u_f = u_{f0} \operatorname{sgn}(Z_3) \tag{4.84}$$

The control structure is similar to the structure of one presented in figure 4.3. Specific features of various types of synchronous motor find the reflection in a choice of values of amplitudes of discontinuous voltages u_{d0}, u_{q0}, u_{f0}. The difference consists in replacing the control loop of angular speed Ω with the one of the stator current component i_q (figure 4.9).

Thus, as well as in the previous case, the results of the first design step are:

- Physically realized control of an excitation current (4.84);
- Allocation in space of two-dimensional fictitious control U an area of the admissible controls U^* providing sliding mode existence.

As shown in section 3.1.3 the control design for the semiconductor power converter feeding the synchronous motor is accomplished in the second step and based on the control designed on the first step for a fictitious control vector $U^T = (u_d, u_q)$ and the chosen values of control amplitude. In this case it completely coincides with the control designed in section 4.1.3 for the various types of the semiconductor power converter. The difference will be only the amplitude of the voltage of the feeding semiconductor power converter. Control block diagrams for synchronous motor, fed from various semiconductor power converters, coinciding with the results in figure 4.6 and 4.7. The difference of the control presented in figure 4.9 is that instead of the angular speed control loop with the references of angular speed Ω_z and of its

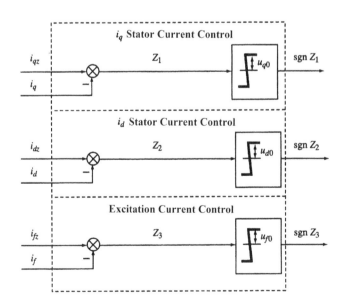

Figure 4.9 Synchronous motor inner control structure

derivative $d\Omega_z/dt$, a control loop of a component of stator current with its reference i_{qz} is used.

4.3 STATIC MODES OPTIMIZATION

4.3.1 Problem statement

As mentioned before, the synchronous motor, in fulfillment of the three-phase condition and depending on the excitation type, has a three-dimensional or a two-dimensional control vector. This allows, besides maintaining of the reference angular speed Ω, to provide high technical and economic indicators of the electrical drive. These indicators concern, first of all, the distribution of the reactive power and the use of the active power. The synchronous motor property not only to consume, but also to generate reactive power opens possibilities on optimization of operating modes by this or that quality criterion. E.g. it would be desirable, that by the active load the complex "semiconductor power converter – synchronous motor" consumed the minimum stator current, i.e. had the maximum efficiency concerning energy losses, or had the coefficient $\cos\varphi = 1$, characterizing consumption of reactive power, so that the complex would not consume any reactive power (Leonhard, 2001), (Boldea & Nasar, 2005), (Bose, 2002), (Mohan et al, 2003).

The possibility of optimization of the synchronous motor mode, i.e. maintaining the formulated above performance requirements, in the frame of the offered control in section 4.1, will be presented below (Ryvkin, 1995). The salient-pole permanent magnet synchronous motors will be used in our presentation. According to the supposed sliding mode control for achieving these requirements there is only one degree of freedom – the reference of a stator current component i_{dz}. In this case, the above mentioned requirements must be transformed in requirements on the reference of the stator current component i_d.

Through the analysis of possibility of reference i_{dz} formatting providing fulfillment of above mentioned requirements, it is assumed that a synchronous motor behavior is completely described by the first harmonic. In this case the static synchronous motor mode is described in a rotating coordinate system by the following equation system:

$$
\begin{cases}
i_s e \sin\varphi = (L_d i_d^2 + L_q i_q^2 + \Psi_f i_d)\omega_z, \\
i_s e \cos\varphi = r i_s + \omega_z M_{el}, \\
M_{el} = M, \\
M_{el} = [\Psi_f - (L_q - L_d)i_d]i_q, \\
i_s = \sqrt{i_d^2 + i_q^2}
\end{cases}
\tag{4.85}
$$

where i_s is the stator current vector module and e is the stator voltage vector one.

The problem consists in formatting based on known parameters of the synchronous motor and a measured component of the stator current component i_q, so that a reference of the stator current component i_{dz}, fulfills one of the above mentioned power requirements taking place.

4.3.2 Keeping maximum efficiency and minimum stator current

Without steel and mechanical losses the second equation (4.85) is the balance equation of active power for the synchronous motor. In this case, the efficiency η can be defined as follows

$$\eta = \frac{\Omega M_{el}}{ri_s + \Omega M_{el}} \tag{4.86}$$

The primary goal of electrical drive control is the mechanical variable control, solving with minimum consumption of electric power. In this case the optimization problem of technical and economic indicators of the electrical drive can be formulated as a problem of maximization of efficiency by constant angular speed and stator current module.

Let us emphasize that the given problem is equivalent to the maximization problem of torque M_{el} developed by the synchronous motor by the fixed module of stator current i_s. For convenience, the solution of this problem of the maximum searching is tried in the polar rotating coordinate system. In this case the electromagnetic torque expression M_{el} is

$$M_{el} = \Psi_f i_s \sin \alpha + 2(L_d - L_q)i_s^2 \sin 2\alpha \tag{4.87}$$

where α is an angle between a stator current vector and the axis d of the rotating coordinate system.

The conditional extreme problem is replaced with an unconditional one on the angle α. From the analysis of the electromagnetic torque M_{el} expression (4.87), taking into account that in the salient-pole synchronous motor condition $L_d > L_q$ (Leonhard, 2001), (Boldea & Nasar, 2005), (Pahman & Zhou, 2000) is always fair, it follows that if the excitation flux $\Psi_f > 0$, the required maximum exists and lies in the first quadrant. The condition of maximum efficiency η, or in other words, a minimum of the stator current module i_s based on (4.87) is

$$\frac{d\eta}{dt} = \frac{dM_{el}}{dt} = \Psi_f i_s \cos \alpha + (L_d - L_q)i_s^2 \cos 2\alpha = 0 \tag{4.88}$$

From this it follows that optimum angle α_{onm} is equal to

$$\alpha_{onm} = \arccos \frac{\Psi_f - \sqrt{\Psi_f^2 + 8(L_q - L_d)^2 i_q^2}}{4(L_q - L_d)} \tag{4.89}$$

Since on problem statement the reference of the stator current component i_d, providing fulfilling above mention requirement should be formed in the close loop by using the information about the stator current component i_q, therefore the received condition of optimality (4.88) should be rewritten in terms of a stator current components

$$\Psi_f i_d + (L_q - L_d)(i_d^2 - i_q^2) = 0 \tag{4.90}$$

From analysis of the electromagnetic torque M_{el} expression (4.85), it follows that the optimum maintenance probably occurs in cases when the sign of the stator current component i_d and that of the excitation flux Ψ_f coincide, the sign of the electromagnetic torque being defined by the sign of the stator current component i_q. Taking into account these remarks it is possible to select, from the two possible solutions of the quadratic equation (4.90), the valid one:

$$i_d = \frac{-\Psi_f + \sqrt{\Psi_f^2 + 4(L_q - L_d)^2 i_q^2}}{2(L_q - L_d)} \qquad (4.91)$$

This solution used to form the reference of the stator current component i_d provides the synchronous motor operation in an optimum mode.

In this case the efficiency reaches the greatest possible value and power losses have their minimum possible value by the given operating mode, defined by the formula:

$$r i_s^2 = r \frac{\Psi_f^2 - \Psi \sqrt{\Psi_f^2 + 4(L_q - L_d)^2 i_q^2} + 4(L_q - L_d)^2 i_q^2}{2(L_q - L_d)^2} \qquad (4.92)$$

4.3.3 Keeping $\cos\varphi = 1$

As it is well known (Leonhard, 2001), (Boldea & Nasar, 2005), (Bose, 2002), (Mohan et al, 2003), (Szentirmai, 2000), (Blaabjerg et al, 2010), (Ohashi, 2010) in a modern economy, problems of rational economic production and consumption of electric power occupy the first place. One of the major factors influencing profitability on the use of power is the reduction of reactive power by a consumed network. The presence of reactive power in a network leads to power stations appearing loaded earlier, and then active loading is reached as planned. Increasing the power factor characterizing a ratio of active and reactive power only by 0.01, in a Russian power supply system, would be the equivalent to producing an additional 500 million kilowatt-hours. It also would allow saving on electrical hardware and labor costs. All elements of electric equipment of three-phase stations and substations (generators, transformers, wires, cables, disconnecting equipment, measuring devices, etc.) are calculated at full power, i.e. they have larger dimensions than those that would be obtained if only active power were transferred. They can be more economical when they receive only active power, because they can be made smaller.

Wide use of various switch inverters in current consumption systems (pumps, fans, compressors, electric drives, power supplies systems, household electric and electronic devices, etc.) has aggravated the problem specified above even more, because these devices, even if they have some known advantages, have a low power factor (0.5–0.7) and high level of harmonics of a current consumed from a network (more than 30%).

A traditional solution to this problem is based on the inclusion in a network structure, cosine transformers and/or synchronous compensators for reactive power, which naturally leads to significant complications and increases in the price of the electrical supply system and does not guarantee a high quality solution because of the impossibility of all consumers to account for reactive power.

On the other hand, there is an alternative "control" possibility to solve the problem. In this case the control of corresponding semiconductor power converters must be designed taking into account the power consumptions requirements. It is necessary to notice that the supply quality requirements and the power consumption requirements (standard IEEE-519, MEK 555, GOST 13109), are now tougher, and the European market is supposed to certify production only according to ISO 9002 norms.

From a power consumption viewpoint, the most preferable operating mode by far is the operating mode without consuming reactive power. The condition to maintain such operating mode is to keep $\cos\varphi = 1$. For the synchronous motor described by (4.85), this condition could be written as

$$L_d i_d^2 + L_q i_q^2 + \Psi_f i_d = 0 \tag{4.93}$$

As it follows from the quadratic equation (4.93) in the static mode the stator current components i_d and i_q are related by following expressions

$$i_{d1} = \frac{-\Psi_f + \sqrt{\Psi_f^2 + 4L_d L_q i_q^2}}{2L_q} \tag{4.94}$$

$$i_{d2} = \frac{-\Psi_f - \sqrt{\Psi_f^2 + 4L_d L_q i_q^2}}{2L_q} \tag{4.95}$$

Proceeding according to the electrical power reasons stated in the previous section, it is convenient to use the stator current i_d expression (4.94) by solving the problem. Because, in this case, the fulfillment of condition $\cos\varphi = 1$ is provided by a smaller value of the stator current component i_d than by one from the expression (4.95), i.e. it helps us to achieve better efficiency. However, functional dependence (4.94) is defined only by $|i_q| \in I_q = \left[0, \Psi_f/2\sqrt{L_d L_q}\right]$ and it is necessary to predetermine the formatting law of the current component on i_d by $|i_q| \notin I_q$. E.g. it is possible to demand the best or most efficient values of $\cos\varphi$, $|i_q| \notin I_q$. Unfortunately, such approach does not allow getting analytical dependence $i_d = f(i_q)$, which could be used in the i_d control loop.

For the purpose of reception of analytical dependence between components of the stator current i_d and i_q by $|i_q| \notin I_q$, the following formatting law of the reference of the current component i_d is offered by $|i_q| \notin I_q$. The i_d reference is constant in this interval and it is equal to the value fulfilling the condition $\cos\varphi = 1$, which provides as much as possible achievable value of electromagnetic torque M_{el}. In this case, the equation connecting variables i_d and M_{el} could be received from (4.85) taking into account that the load torque M is positive, condition $\cos\varphi = 1$ is fulfilled, and components of current i_d and i_q are related by the following expression:

$$i_q = \frac{\sqrt{-\Psi_f i_d - L_d i_d^2}}{L_d} \tag{4.96}$$

which can also look like this:

$$L_d(L_q - L_d)^2 i_d^4 - (L_q - Lx_d)\Psi_f(3L_d - L_q)i_d^3$$
$$+ \Psi_f(3L_d - 2L_q)i_d^2 + \Psi_f^3 i_d + x_q M^2 = 0 \tag{4.97}$$

This is a fourth degree real equation with positive factors of variable i_d. Unfortunately, it is not explicitly solvable. Based on the Descartes signs rule (Korn & Korn, 2000), it is possible to assert that equation (4.97) can have real negative roots or complex roots with a negative real part. Only the negative real roots, which the dependence $i_d = f(M)$ has as optimal, are interesting.

Let us consider the inverse function $M = f(i_d)$ and try to allocate areas of single-valued dependence M and i_d. For this purpose the function (4.97) was investigated on an extreme in the area $I_d^* | i_d | \in I_d^* = [-\Psi_f/x_d, 0]$, which corresponds to performance maintenance of the condition $\cos \varphi = 1$

$$\frac{dM}{di_d} = \frac{4(L_q - L_d)L_d i_d^2 - (5L_d - 3L_q)\Psi_f i_d - \Psi_f^2}{2L_q\sqrt{-L_d i_d^2 - \Psi_f i_d}} = 0 \tag{4.98}$$

At $M \neq 0$ and taking into account (4.96) that the denominator in (4.98) is distinct from zero, the extreme condition takes the form:

$$4(L_q - L_d)L_d i_d^2 - (5L_d - 3L_q)\Psi_f i_d - \Psi_f^2 = 0 \tag{4.99}$$

This equation has two solutions for i_d, namely:

$$i_{d1} = \frac{(5L_d - 3L_q)\Psi_f - \sqrt{D}}{8(L_q - L_d)L_d} \tag{4.100}$$

$$i_{d2} = \frac{(5L_d - 3L_q)\Psi_f + \sqrt{D}}{8(L_q - L_d)L_d} \tag{4.101}$$

where D is the equation (4.99) discriminant and

$$D = (5L_d - 3L_q)^2\Psi_f^2 + 16(L_q - L_d)L_d\Psi_f^2 > 0 \tag{4.102}$$

It is easy to show that only $i_{d1} \in I_d^*$. This fact specifies ambiguity of dependence $M = f(i_d)$ at $|i_d| \in I_d^*$. On the borders of this interval $M = 0$, and in it, when $i_d = i_{d1}$ the electromagnetic torque reaches the maximum value, fulfilling the condition $\cos\varphi = 1$. To avoid ambiguity, it is enough to choose i_d from the range of I_d^* fulfilling the following equations $I_{d1}^* = [-\Psi_f/L_d, i_{d1}]$ or $I_{d2}^* = [i_{d1}, 0]I_{d1}^* \cup I_{d2}^* = I_d^*$, $I_{d1}^* \cap I_{d2}^* = i_{d1}$.

From the power viewpoint it is expedient to use the functional dependence $M_{el} = f(i_d)$ on the range I_{d2}^*. Therefore the formatting law of the i_d reference looks like this:

$$
i_d = \begin{cases}
\dfrac{-\Psi_f + \sqrt{\Psi_f^2 + 4L_d L_q i_q^2}}{2L_q}, & if\ |i_q| \in I_q^* \\[4ex]
\dfrac{(5L_d - 3L_q)\Psi_f - \sqrt{(5L_d - 3L_q)^2 \Psi_f^2 + 16(L_q - L_d)L_d \Psi_f^2}}{8(L_q - L_d)L_d}, & if\ |i_q| \notin I_q^*
\end{cases}
$$

$$(4.103)$$

where $I_q^* = [0, i_{q1}]$, $i_{q1} = (1/x_q)\sqrt{-x_d i_{d1}^2 - \Psi_f i_{d1}}$. In this case, if the maximum attainable value of electromagnetic torque, when the condition $\cos\varphi = 1$ exceeds the value of load torque, the static mode is supported.

4.3.4 Realization of the offered dependencies

The above obtained dependence of the stator current component i_d from the stator current component i_q allows providing demanded synchronous motor static operating modes by using as the source information about the phase currents. It is obvious that it is possible to provide permanent realization of these dependencies. However it would lead to necessity of additional researches of the offered in section 4.1 controls. As by the control design it was supposed that the angular speed Ω and the stator current component i_d are independent variables. The reference of the stator current component i_{dz}, in the given section, providing demanded power operating mode is formed by a feedback principle. In this case the reference of the stator current component i_{dz} naturally depends on the angular speed of the synchronous motor Ω, i.e. is not an independent variable. If it were possible in the frame of the offered in section 4.1, controls not only would fulfill the mechanical variable requirements, but also the power requirement offered in sections 4.3.2, 4.3.3, and rewritten as the reference of the stator current component i_{dz}, it is necessary to make a close loop of reference, forming the current component independent. In particular, it can be done by including a filter with the time constant exceeding the electromechanical transient time of the synchronous motor in the loop of reference forming of the stator current component i_{dz}. In this case the control loop of angular speed Ω and the loop of the reference forming of the stator current component i_{dz} have essentially various transient rates. Also, the reference of the stator current component i_d is always formed on values of a quasi-static mode. It is necessary to notice that the way of division of movement specified above is widely used by automatic control system design (Tikhonov, 1952), (Kokotovic, 1976), (Krstic, 1995).

For stability research of the loop of the reference forming of the stator current component i_d the differential equation system of the synchronous motor, added to the differential equation of the loop of reference forming, is used.

$$\frac{di_d}{dt} = \frac{1}{L_d}(-ri_d + L_q\omega i_q) + \frac{1}{L_d}u_d,$$

$$\frac{di_q}{dt} = \frac{1}{L_q}(-ri_q - L_d\omega i_d - \Psi_f\omega) + \frac{1}{L_q}u_q,$$

$$\frac{d\omega}{dt} = \frac{3p}{2J}(M_{el} - M), M_{el} = [\Psi_f - (L_q - L_d)i_d]i_q,$$ (4.104)

$$\frac{di_{dz}}{dt} = \frac{1}{T}(i_d^* - i_{dz})$$

where t is the time; T is the time constant of the filter, i_d^* is the reference formed according to (4.91) or (4.103) depending on static mode requirements and arriving on the filter.

The system (4.104) could be rewritten using a new time constant $\tau = t/T$ and the new parameter $\mu = 1/T$. In accordance with the problem statement, the time constant T is large, so the parameter μ is small, and it is possible to take advantage of the properties of singular systems (Tikhonov, 1952), (Kokotovic, 1976), (Krstic, 1995). In the limiting case that $\mu \to 0$ the system (4.104) looks like:

$$0 = \frac{1}{L_d}(-ri_d + L_q\omega i_q) + \frac{1}{L_d}u_d,$$

$$0 = \frac{1}{L_q}(-ri_q - L_d\omega i_d - \Psi_f\omega) + \frac{1}{L_q}u_q,$$

$$0 = \frac{3p}{2J}(M_{el} - M), \quad M_{el} = [\Psi_f - (L_q - L_d)i_d]i_q,$$ (4.105)

$$\frac{di_{dz}}{dt} = i_d^* - i_{dz}$$

The stability of this system will be analyzed using the second method of Lyapunov and a Lyapunov function like this:

$$V = \frac{1}{2}(i_{d0} - i_{dz})^2$$ (4.106)

where i_{d0} is the steady-state value of the stator current component i_d, defined according to a researched power mode.

If the derivative of function (4.106)

$$W = -(i_{d0} - i_{dz})(-i_{dz} + i_d^*)$$ (4.107)

is negative definite on system trajectories (4.105), the loop of the reference forming of the stator current component will be stable. For this condition fulfilling it is necessary, that both factors in (4.107) had identical signs, i.e. at $i_{dz} > i_{d0}, (-i_{dz} + i_d^*) \leq 0$ and on the contrary.

For maximum efficiency the second factor looks like:

$$-i_{dz} + \frac{\Psi_f - \sqrt{\Psi_f^2 + 4(L_q - L_d)^2 \left[\frac{M}{\Psi_f - (L_q - L_{dq})i_{dz}}\right]^2}}{2(L_q - L_d)} \tag{4.108}$$

After transformations, the following expression was obtained:

$$-(L_q - L_d)^3 i_d^4 + 3(L_q - L_d)^2 \Psi_f i_d^3 - 3\Psi_f^2(L_q - L_d)i_d^2 + \Psi_f^3 i_d - M \tag{4.109}$$

which must be greater or less than zero.

This expression coincides with the expression standing at the left in the functional dependence $i_d = f(M)$, providing the maximum efficiency. This is a fourth degree real equation. According to Descartes' rule (Korn & Korn, 2000), it has one positive root i_{d0} by each concrete value of the load torque. It is obvious that in this case when $i_{d0} > i_{dz}$, the expression (4.107) is positive, and it is negative when $i_{d0} < i_{dz}$, i.e. the loop of the reference forming of the stator current component is stable.

In case of no reactive power consumption mode, i.e. $\cos\varphi = 1$, stability of the loop of reference forming the stator current component, it is necessary to search only on the I_q^* area. Out of it, the reference of the stator current component i_{dz}, according to (4.103), is a constant, i.e. independent, and the control conditions used in section 4.1 are met. For the stator current component from the area I_q^* the second factor in expression (4.107) taking into account (4.103) looks like:

$$-i_{dz} + \frac{-\Psi_f + \sqrt{\Psi_f^2 + 4L_q L_d i_q^{*2}}}{2L_q} \tag{4.110}$$

After transformations the following expression is obtained:

$$L_d(L_q - L_d)^2 i_d^4 - (L_q - Lx_d)\Psi_f(3L_d - L_q)i_d^3 + \Psi_f(3L_d - 2L_q)i_d^2 + \Psi_f^3 i_d + x_q M^2 \tag{4.111}$$

which must be less or greater than zero.

This expression coincides with the expression standing on the left of the functional dependence $i_d = f(M)$ (4.97), providing $\cos\varphi = 1$. This is a fourth degree real equation. According to Descartes' rule (Korn & Korn, 2000) it has one positive root i_{d0} for each concrete value of the load torque. It is obvious that in this case, when $i_{d0} > i_{dz}$, the expression (4.111) is positive, and when $i_{d0} < i_{dz}$, it is negative, i.e. the loop of the reference forming of the stator current component i_d is stable.

Proceeding from all above stated, it is possible to draw a conclusion that the loop of the reference forming of the stator current component i_d is stable in both cases, i.e. the use of the laws of the forming reference i_{dz} does not lead to infringement of system stability.

4.3.5 Using control task $i_{dz} = 0$

The above received algorithms of the reference forming of the stator current component i_d allow to support precisely $\cos \varphi = 1$ and to provide the maximum value of efficiency. However their hardware realization is complex enough.

In practice, to simplify realization, it is often convenient to make the support value $\cos \varphi = 1$, so that efficiency is near its optimum values. In this case the reference of the stator current component i_d is formed as $i_{dz} = 0$. Then the value of the angle φ is defined by the expression $tg\varphi = L_q i_q \omega_z / (r i_q + \omega_z \Psi_f)$ and it is always positive when $i_q > 0$. When i_q increases, φ changes from 0 ($\cos \varphi = 1$) to $arctg L_q \omega_z / r$. Power losses in this case differ from the optimum by an amount determined by the following formula:

$$\Delta P = [\Psi_f^4 + \Psi_f^3 \sqrt{\Psi_f^2 + 4(L_q - L_d)^2 i_q^2} + 3\Psi_f^2 (L_q - L_d)^2 i_q^2$$
$$- \Psi_f (L_q - L_d)^2 i_q^2 \sqrt{\Psi_f^2 + 4(L_q - L_d)^2 i_q^2} - 2(L_q - L_d)^4 i_q^4]/[2\Psi_f^2 (L_q - L_d)^2]$$

$$(4.112)$$

where i_q is the stator current component value, corresponding to optimum power losses.

REFERENCES

Blaabjerg, F., Iov, F., Kerekes, T. and Teodorescu, R. "*Trends in power electronics and control of renewable energy systems*". Proc. 14th International Power Electronics and Motion Control Conference, EPE-PEMC 2010, Ohrid, Republic of Macedonia, 2010, pp. K-1–K-19.

Boldea, I. and Nasar, S.A. *Electric drives*, 2nd ed. CRC Press, 2005. 544 p.

Bose, B.K. *Modern power electronics and AC drives*. New Jersey: Prentice Hall, 2002. 711 p.

Holz, J. "*Pulsewidth modulation for electronic power conversion*". Proc. of the IEEE, 1994, vol. 82, no. 8, pp. 1194–1213.

Kokotovic, P.V., O'Malley, R.B. and Sannuti, P. "*Singular perturbation and reduction in control theory*". Automatica, 1976, no. 12, pp. 123–132.

Korn, G.A. and Korn, T.M. *Mathematical handbook for scientists and engineers*. USA: Dover Publications, 2000. 1152 p.

Krstic, M., Kanellakopoulos, I. and Kokotovic, P. *Nonlinear and Adaptive Control Design*. New York: Wiley, 1995. 563 p.

Kuerker, O. Modulation techniques. *Modern Electrical Drives*. Dordrecht, Boston, London: Kluwer Academic Publishers, 2000, pp. 289–310.

Leonhard, W. *Control of electrical drives*. Berlin: Springer-Verlag, 2001. 460 p.

Mohan, N., Underland, T.M. and Robbins, W.P. *Power electronics: converters, applications and design*. 3rd ed. New York: John Wiley & Son Inc., 2003. 824 p.

Ohashi, H. "*Role of green electronics in low carbonated society toward 2030*". Proc. 14th International Power Electronics and Motion Control Conference, EPE–PEMC 2010, Ohrid, Republic of Macedonia, 2010, pp. K-20–K-25.

Pahman, M.A. and Zhou, P. *Interior permanent magnet motors. Modern Electrical Drives*. Dordrecht, Boston, London: Kluwer Academic Publishers, 2000, pp. 115–140.

Ryvkin, S. *"Permanent magnet synchronous motor with sliding mode control"*. Proc. the 4th European Conference on Power Electronics and Applications, Florence, Italy, 1991, pp. 382–387.

Ryvkin, S. *"Sliding mode control of a synchronous reluctance motor"*. Proc. the International Conference on Recent Advances in Mechatronics, ICRAM'95. Istanbul, Turkey, 1995, pp. 580–583.

Ryvkin, S. *"Using sliding mode in synchronous motor control"*. Technical electrical dynamics, 1982, no. 4, pp. 58–63 (In Russian).

Ryvkin, S. *Optimization of static modes of a synchronous motor. Discontinuous control systems with sliding mode*. Moscow: Institute of Control Sciences, 1983, pp. 36–43 (in Russian).

Szentirmai, L. *Considerations on industrial drives. Modern Electrical Drives*. Dordrecht, Boston, London: Kluwer Academic Publishers, 2000, pp. 687–722.

Tikhonov, N. *"Systems of differential equations with a small parameter multiplying derivations"*. Mathematicheskii Sbornik, 1952, vol. 73, no. 31, pp. 575–586 (In Russian).

Chapter 5

Multidimensional switching regularization

5.1 FEATURES OF REAL SLIDING MODE

As specified in section 2.1, systems controlled using sliding mode, i.e. with control switched on the switching surfaces, possess high speed and small sensitivity to changes of parameters and external disturbances. However, when the system is in multidimensional sliding mode there is not just one but many discontinuous controls, and a problem arises with the sequence of their switching components. The existence conditions of a sliding mode, as it is known, look like inequalities and ambiguously define an algorithm of multidimensional discontinuous control. There are various possibilities for control design, which provide existence of sliding mode. A variety of sliding mode controls lead to a mixture of switching sequences. Moreover, the switching frequency is achieved using modern power switches: insulated gate bipolar transistor (IGBT), metal oxide semiconductor field-effect transistor (MOSFET), Gate Turn-Off (GTO) thyristor etc. Even if their switching frequency can reach tens or even hundreds of kHz, it is nonetheless finite. Therefore, the resulting sliding movement is carried out in the vicinities δ of the crossing of sliding surfaces. Such sliding movement is named real sliding motion or real sliding mode.

In this sliding motion, dynamic processes of a limit cycle establishment take place. However, the real sliding motion is sensible to arbitrary small variations of system parameters and has bifurcation changes in the switching sequence of discontinuous control components, i.e. the transition from one limit cycle to another. Such processes are characterized by a chaotic multidimensional control switching (Nagy, 1994), (Suetz et al, 1996), (Holz, 1994). The control error thus does not exceed defined values, e.g. one defined by the value of the switch hysteresis forming a control component. However the change dynamics and a chaotic switching lead to deterioration of electrical drive technical and economic indicators because of sharp increase of switching losses, admissible excess from a position of electromagnetic compatibility of frequency (>10 kHz), occurrences of acoustic noise (1–2 kHz) at the expense of influence of Lorentz force on the motor ferromagnetic materials (Holz, 1994).

It is obvious that maintaining regular switching in real sliding mode allows eliminating the drawbacks mentioned above and, as a consequence, it improves the electrical drive technical and economic indicators. Some variants of such discontinuous controls will be presented below.

5.2 SWITCHING LOSS MINIMIZED CONTROL FOR VSI

5.2.1 Analysis of PWM laws

From the viewpoint of the DC feed voltage control, real sliding mode is characterized by the automatic PWM organization in a closed loop. This is PWM of the second sort, i.e. with feedback, which is inherent in all relay systems (Kuntsevich & Chehovoy, 1970). In this case a possible way of regulating PWM organization could be to make its properties equivalent to PWM of the first sort, i.e. the feedforward or programmed one. The latter is formed, as a programmed control of the semiconductor power converter that allows to provide demanded frequency and power properties of a complex "semiconductor power converter – synchronous motor" by the control problem solving for the electrical drive.

As mentioned in section 1.2, the use of PWM of the first sort for a synchronous drive control is the basis of one of the classical control design approaches called one-loop decomposition control. It consists in breaking up the control problem into two separate problems:

- Synchronous motor control;
- Semiconductor power converter control.

Thus the problem of synchronous motor control is identical to the primary and main control one, i.e. a mechanical coordinates control problem, involving often the angular speed $\Omega(t)$, in a combination to maintaining the technical and economic indicators of the electrical drive. The problem of semiconductor power converter control is secondary in relation to the synchronous motor control. It consists in forming such three-phase voltage at the semiconductor power converter output fed the synchronous motor, that is necessary for the solving the main control problem. Incidentally, this is an independent problem of feedforward or direct control and could be solved separately.

Nowadays, the majority of existing controls for a semiconductor power converter, and, in particular for a VSI, as specified in section 1.1.2, are based on the PWM principle. Its main feature, as a program or feedforward control, is that the indicators characterizing modulation, such as frequency and pulse ratio, are formed for VSI from the outside.

The aim of VSI control is maintaining the demanded average output VSI voltage provided the main control problem is solved, i.e. the mechanical coordinate control. Since real physical VSI controls are switching control commands, the problem of VSI control in case of modulation use consists in formation of switch control commands p_j so that the average value of the output voltage during the modulation period was equal to the demanded value of output voltage. It is natural that the realized average values of the output voltage lie in a hexagon formed by instant values an output voltage vector (figure 1.5). However, the same average value can be obtained with use of an interim modulation period to average the various instantaneous values of the output voltage vector, i.e. different algorithms and various VSI switch controls. Thus, in the VSI control design based on the requirements to ensure the required average value for the time interval of averaging the output voltage, there is ambiguity in the solution due to control redundancy. This ambiguity can be removed by using additional VSI work requirements, e.g.:

- Minimization of the electric power losses;
- Minimization of the number of switchings during the modulation period;
- Maximization of an achievable time interval between switching of one phase power switches;
- Full use of input constant voltage U_{in};
- Maintenance of symmetric control with phases VSI, etc.

When designing PWM the main condition is that the modulation period is a pre-determined value, and it is an interval of output voltage averaging. The VSI switches control providing reception of demanded output voltage includes two independent components (Zinoviev, 2005), (Trzynzdlowsky & Legowski, 1994):

The modulation law of the phase switches, defining the switch-on interval duration of switch on the modulation period, i.e. switch-on on the line (U_{in}) or (0), accordingly with reference voltage value;

The switching law of phase switches, defining the switching sequence of the phases switches on the modulation period.

In this case, each VSI phase output voltage represents a sequence of squared impulses of various durations, whose amplitude is equal to feed voltage U_{in}. A sequence of these impulses, being averaged, due to the filtering properties of the load, forms a phase output voltage on the load, which must be controlled. Thus, it is necessary to consider the dual character of an output voltage vector at VSI analysis. It can be regarded either as an instantaneous output voltage caused by the power switching or as the average output voltage on the VSI modulation period.

For the design of the load average voltage, it is convenient to use the three instant vector of the output voltage, one of which is zero, and the other two are the closest to demanded average value. Such choice of the output voltage vector provides minimization of a harmonic component in the average output voltage.

When we implement a required average voltage vector $U_{ref}(t)$ on the modulation period with the use of three instant load voltage vectors U_0, U_1 and U_2 (figure 1.5), the modulation law is uniquely defined as follows (Holz, 1994), (Zinoviev, 2005), (Pfaff & Wick, 1983).

$$\mu_o = 1 - \chi \cos \varphi; \quad \mu_1 = -\chi \sin(\varphi - \pi/6); \quad \mu_2 = \chi \sin(\varphi + \pi/6) \tag{5.1}$$

where μ_i represents the ratio of the modulation period during which each of three used vectors of instant voltage U_0, U_1 and U_2, is switched-on, $\sum_{i=1}^{3} \mu_i = 1$; χ is the amplitude reference of the relative average voltage vector, $\chi = \sqrt{3}A/U_{in}$, A is the reference value of the average voltage vector U_{ref} amplitude, φ is the reference value of the average voltage vector U_{ref} angular position, defined as a deviation of the average voltage vector from a bisector of the angle formed by the used instant voltage vectors (figure 1.5).

In contrast to the design of the modulation law that is unequivocally defined by the reference average voltage vector, the switching law is not strictly deterministic and gives some degree of freedom in choosing the switching sequence of the phases switches in the modulation period. This freedom can be used to satisfy the VSI additional working requirements specified above. For instance, minimization of switching losses, whose value increases as PWM frequency reaches hundreds of kHz. This problem will be analyzed and solved below.

As it was specified above, one of features of VSI functioning is that the instantaneous zero voltage value can be received in two ways. Either all switches on (i.e. the switch command combination is described as (111)) or off (i.e. the combination is described as (000)). For appropriate distinction of these switch status combinations to a zero-vector of output voltage U_0, the additional top index "−", i.e. U_0^+, and "+", i.e. U_0^-, will be respectively used.

The dominating factor of switching losses minimization is the number of power switchings on the modulation period. It happens according to the following cyclic switching laws:

$$U_0^- \Rightarrow U_1 \Rightarrow U_2 \tag{5.2}$$

$$U_0^- \Rightarrow U_2 \Rightarrow U_1 \tag{5.3}$$

$$U_0^+ \Rightarrow U_1 \Rightarrow U_2 \tag{5.4}$$

$$U_0^+ \Rightarrow U_2 \Rightarrow U_1 \tag{5.5}$$

$$U_0^- \Rightarrow U_1 \Rightarrow U_2 \Rightarrow U_1 \tag{5.6}$$

$$U_0^+ \Rightarrow U_2 \Rightarrow U_1 \Rightarrow U_2 \tag{5.7}$$

$$U_0^- \Rightarrow U_1 \Rightarrow U_2 \Rightarrow U_0^+ \Rightarrow U_2 \Rightarrow U_1 \tag{5.8}$$

It is necessary to note that the following features come as a result of the above switching laws:

- If in the switching law any instantaneous output vector voltage occurs twice, it means that the total time spent in this vector position is determined by the modulation law, and its division into fraction of the time corresponding to each of the two visits, carried out on the basis of additional requirements. For example, in this case, from the symmetry requirement of the power switches, it amounts for each visit to half the estimated time;
- In the implementation of the switching laws (5.2)–(5.5), one of the phases during the entire modulation period remains connected to either of two input voltage buses. Phase selection and bus input voltage are defined by the implemented switching law;
- Four switchings on the modulation period correspond to switching laws (5.2)–(5.7), and six switchings correspond to the switching law (5.8);
- A specific feature of switching laws (5.2)–(5.5) is the simultaneous switching of switches of two phases at transition from status U_0^- to U_2, from U_0^+ to U_1 and back. As opposed to above mentioned switching laws, the other ones correspond a serial switching of power switches of various phases;

Given all the above and taking into account the additional requirement of symmetry control, the basis for the PWM algorithm design with minimum switching losses are switching laws (5.6), (5.7) and (5.8), which will be called from now on, α-, β- and γ-switching laws, respectively.

Thus we will note the following fact: all variety of known algorithms of vector PWM is finally based on the use of one of the above resulting switching laws. For example, the widely known PWM with sawtooth voltage (Mohan et al, 2003) is based on the same switching sequence as the γ-switching law. The difference consists in the splitting time implementation of the zero vectors in one cycle. The switching law using the PWM sawtooth voltage usually has unequal intervals of implementation of the zero vector output voltage. The inequality of the specified time intervals is caused by way of their definition, which is based on comparison of three phases of sinusoidal output voltage of the fundamental frequency with sawtooth voltage of much higher modulating frequency. Points of intersection of corresponding sinusoids with a sawtooth signal also define inclusion times and switch disconnections. In this sense we can speak of a "centered" PWM sawtooth voltage, which has diagrams of switching phase voltages symmetric to each other. (In the case of conventional PWM with sawtooth voltage this takes place when the voltage of one of the phases becomes zero, and consequently, the other two phase voltages have opposite signs). In the following arguments with the additional requirements of symmetry control when considering γ-switching law, we always keep in mind "symmetric PWM" with equal time of implementation of a zero-vector.

5.2.2 Comparative analysis of switching laws from the switching losses viewpoint

As it was specified above, switching losses are caused first of all by the amount of switching on the modulation period, and at their constancy character of a deviation of instant values of the output voltage (current) from desirable average value. From these positions, switching laws (5.2)–(5.7) have four switchings on the period and for their comparison on switching losses it is necessary to use the second criterion. As to the γ-switching law, it has six switchings on the period that, at first sight does not allow carrying it to preferable candidates on the switching law with minimum switching losses. However, it has a conclusive advantage: the symmetric structure, allowing using all available possibilities of VSI switches control, does a desirable estimation on its switching losses. To maintain on γ-switching law of the same entry conditions on the number of switchings in the standard period, as well as at other switching laws, its modulation period should be equal to one and a half the standard modulation period. In this case, on the standard period of modulation, all switching laws will have four switchings.

As it was mentioned above, when using high-frequency power switches most losses are caused by switching losses. These losses are caused, first, by the number of switchings on the period, and, second, by the character of deviation of instant values of output voltage (current) from the desirable average value. The analysis must take into account that by using high-frequency power switches, the modulation period T is small in comparison to the period of the basic VSI output voltage harmonic (Shevtsov et al, 1994), (Ryvkin & Isozimov, 1997). To assess additional switching losses in the load caused by deviations of output voltage, the validity and suitability of use of integrated quadratic criterion of quality of the VSI output voltage is physically obvious:

$$J = \frac{1}{T} \int_0^T N[\Delta I(t)]^2 dt \tag{5.9}$$

where $\Delta I(t)$ is a output current error vector, $\Delta I(t) = I(t) - I_{ref}(t)$, $I(t)$ is the actual value of a VSI output current vector, $I_{ref}(t)$ is a reference value of a VSI output current vector or its average value, N is the normalizing factor. Obviously, the electric motor is usually an inductive VSI load. This criterion has a clear physical meaning: the losses are due to the resistance of the copper windings. Let us calculate the criterion (5.9) to determine the time plot of the vector deviation of the output current from its average value. Taking into account that when using high power switches, the modulation period T is small compared to the period of the fundamental harmonic of output voltage, which remains constant, the time plot of the current error vector represents a set of linear segments, whose direction is determined by unit error vectors, or the ort of voltage error vector $\Delta U_i(t)$, which is the difference between the instantaneous output voltage vector $U_i(t)$ and the reference average voltage vector $U_{ref}(t)$, according to this formula: $\Delta U_i(t) = U_i(t) - U_{ref}(t)$.

To describe the behavior of the current error vector $\Delta I(t)$, we use the orthogonal coordinate system (p, r) with a rigid axis p connected with the average output voltage vector $U_{ref}(t)$. Figure 5.1 shows the vector time plots of a current error vector for all the above stated switching laws.

In accordance with the current depicted vector time plots, the above mentioned criterion of estimating switching losses (5.9) is the sum of integrals on each of the line segments of the time plot:

$$J = N \sum \int_0^{T_i} (\Delta I(t))^2 dt \Big/ \sum T_i, T_i = \mu_i T \qquad (5.10)$$

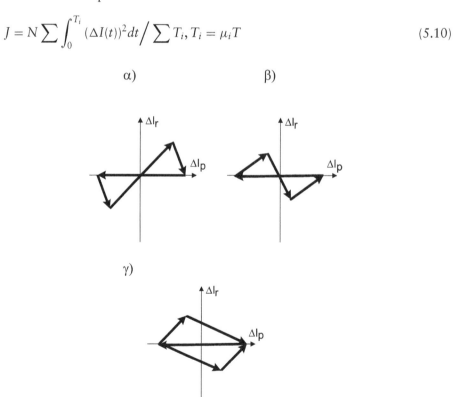

Figure 5.1 Current error vector $\Delta I(t)$ graph with α, β and γ PWM switching laws

where the line segment number corresponds to the number of the output voltage instant vector.

As it is known, the integral value on a line segment is defined by its initial and final conditions. Therefore, taking into account orthogonal components current errors $\Delta I_p(\tau)$ and $\Delta I_r(\tau)$ representation on a line segment, we have:

$$\Delta I_p(\tau) = (\Delta I_{pif} - \Delta I_{pii})t/T_i + \Delta I_{pii},$$
$$\Delta I_r(\tau) = (\Delta I_{rif} - \Delta I_{rii})t/T_i + \Delta I_{rii} \qquad (5.11)$$

where the bottom index has the following structure: the first position is a coordinate direction; the second one represents the number of a line segment of a the vector time plot; the third one shows this value belonging to the beginning (i) or to the end (f) to line segment.

Taking into account (5.10) and (5.11), integrated square-law criterion of VSI output voltage quality (5.9), it is defined by the following algebraic expression:

$$\int_0^{T_i} (\Delta I(t))^2 dt = [(\Delta I_{pif} - \Delta I_{pii})^2 + (\Delta I_{rif} - \Delta I_{rii})^2]T_i/12$$
$$+ [(\Delta I_{pif}/2 - \Delta I_{pii}/2)^2 + (\Delta I_{rif}/2 - \Delta I_{rii}/2)^2]T_i \qquad (5.12)$$

The first expression term is proportional to a square of the length of the i-th line segment of the time plot, and the second one is proportional to the square distance from the line segment middle to the origin of coordinates. Lengths of various line segments for various above mentioned switching laws (5.2)–(5.8) are defined as follows:

For a zero line segment

$$\Delta I_{p0f} - \Delta I_{p0i} = -\chi/T_0, \qquad (5.13)$$
$$\Delta I_{r0f} - \Delta I_{r0i} = 0$$

For a first line segment

$$\Delta I_{p1f} - \Delta I_{p1i} = (2\cos(\varphi + \pi/6)/\sqrt{3} - \chi)T_1, \qquad (5.14)$$
$$\Delta I_{r1f} - \Delta I_{r1i} = -2\sin(\varphi + \pi/6)T_1/\sqrt{3}$$

For a second line segment

$$\Delta I_{p2f} - \Delta I_{p2i} = (2\cos(\varphi - \pi/6)/\sqrt{3} - \chi)T_2, \qquad (5.15)$$
$$\Delta I_{r2f} - \Delta I_{r2i} = -2\sin(\varphi - \pi/6)T_2/\sqrt{3}$$

The position of initial, final and middle points of the i-th line segment is determined from the symmetry of the current error vector $\Delta I(t)$ time plots for α, β and γ PWM switching laws. For switching laws (5.2)–(5.5) for zero average value of the current error ΔI on the modulation period is it necessary to satisfy additional conditions:

$$\sum (\Delta I_{pif} + \Delta I_{pii})T_i = 0, \tag{5.16}$$

$$\sum (\Delta I_{rif} - \Delta I_{rii})T_i = 0$$

Thus, for the switching losses the comparative analysis of all the considered PWM switching laws, the offered criterion of quality (5.9) can be considered a function of the average value of the output voltage on the modulation period. This average value is characterized by an absolute A or relative χ amplitude and an angle φ between an average voltage vector and the angle voltage bisector.

5.2.3   Comparing PWM switching laws – Numerical results

As specified above the α, β and γ switching laws are searched for optimal switching losses. Tables 5.1 and 5.2 present normalized values of proposed criterion for α, β and γ switching laws (valuation reached the maximum value of the criterion).

After comparing numerical values in tables 5.1 and 5.2 it is obvious that by using $\alpha(\beta)$ switching law the maximum value of criterion (5.9) does not exceed 0.444.

Table 5.1 Numerical values of criterion (5.9) for $\alpha(\beta)$ switching laws (For β switching law the φ angle has the opposite sign).

χ/φ	$-\pi/6$	$-\pi/8$	$-\pi/12$	$-\pi/24$	0	$\pi/24$	$\pi/12$	$\pi/8$	$\pi/6$
.999	.096	.092	.240	.387	.444	.392	.270	.148	.096
.875	.240	.169	.225	.309	.362	.359	.315	.263	.240
.750	.369	.282	.289	.333	.375	.394	.391	.377	.369
.625	.438	.363	.352	.372	.401	.425	.436	.439	.438
.500	.429	.378	.363	.371	.389	.407	.420	.427	.429
.375	.342	.315	.306	.308	.316	.327	.335	.340	.342
.250	.205	.195	.191	.192	.194	.198	.202	.204	.205
.125	.066	.065	.064	.065	.065	.066	.066	.066	.066

Table 5.2 Numerical values for criterion (5.9) for γ switching law

χ/φ	$-\pi/6$	$-\pi/8$	$-\pi/12$	$-\pi/24$	0	$\pi/24$	$\pi/12$	$\pi/8$	$\pi/6$
0.999	0.054	0.218	0.563	0.876	1.000	0.876	0.563	0.218	0.054
0.875	0.135	0.234	0.442	0.631	0.706	0.631	0.442	0.234	0.135
0.750	0.207	0.263	0.380	0.486	0.527	0.486	0.380	0.263	0.207
0.625	0.247	0.275	0.334	0.388	0.409	0.388	0.334	0.275	0.247
0.500	0.241	0.254	0.280	0.303	0.313	0.303	0.280	0.254	0.241
0.375	0.192	0.197	0.206	0.214	0.218	0.214	0.206	0.197	0.192
0.250	0.115	0.116	0.118	0.120	0.121	0.120	0.118	0.116	0.115
0.125	0.037	0.037	0.038	0.038	0.038	0.038	0.038	0.037	0.037

Table 5.3 Particular cases of criterion calculation

Vector $U_{ref}(t)$	J_γ/J_α (J_γ/J_β)	Optimum switching law
U_0 or U_1 or U_2	$J_\alpha = J_\beta = J_\gamma = 0$	Any of the three
ort U_1 or ort U_2	$J_\gamma/J_\alpha = J_\gamma/J_\beta = 9/16$	γ
$(U_1 + U_2)/2(\chi = 1, \varphi = 0)$	$J_\gamma/J_\alpha = J_\gamma/J_\beta = 9/4$	α or β
$(\chi = 2/3, \varphi = 0)$	$J_\gamma/J_\alpha = J_\gamma/J_\beta = 9/8$	α or β

For some simple cases the reference output voltage vector $U_{ref}(t)$ realization calculation of criterion does not represent work, and the received results for switching law selection are obvious (table 5.3).

5.2.4   Switching loss minimizing PWM

From the analysis of computing results presented in tables 5.1 and 5.2 it is obvious that for any output voltage, one of the three switching laws vector minimizes the criterion. The design problem of switching loss minimizing PWM consists, in this case, first, in splitting of all realized output voltage space into three areas connected with the switching law providing the chosen criterion (5.9) for minimum (Ryvkin & Isozimov, 1997). Second, it must ensure a smooth interface of various switching laws (Ryvkin et al, 1998).

Figure 5.2 presents the above mentioned splitting in the coordinate system (χ, φ), for the first sector.

Given the optimal partition of realizable output voltage should be extended to an arbitrary location output voltage vector $U_{ref}(t)$ in any of the six sectors defined by the instant VSI output voltage vectors. This problem is due to the symmetry of the instant VSI output voltage vector locations can be easily solved by using logical table 5.4. The renaming the instant VSI output voltages are connected with the angle position of the reference average vector.

Another important problem that must be solved is the coupling of the various optimal switching laws. In the frame of the switching loss minimizing PWM there is a transition from one switching law to another, caused by the change of the reference voltage and/or of DC line voltage. This transition should not lead to a step-up change of the VSI average output voltage and implemented with a minimal number of additional, transient switching.

From the viewpoint of the above requirements for the transition from one switching law to another one, it is preferable to use as cycle initial points null vectors U_0^+ or U_0^-. In this case, the transition from α switching law to β one, or vice versa, depending on the border crossings, would have no additional switchings (crossing the sector border), or two additional switchings (crossing the sector bisector) (figure 1.5). By coupling any of the above mentioned laws and γ one there is a conjugation problem of two current error vector $\Delta I(t)$ time plots with different modulation period. Note that the modulation period for γ switching law is up to 1.5 times greater than $\alpha(\beta)$, due to the transition from $\alpha(\beta)$ to γ switching laws (figure 4.1), to ensure continuity. For the time plot of the initially calculated γ law, the relative time μ_0 must be increased by the value $\mu_0/8$.

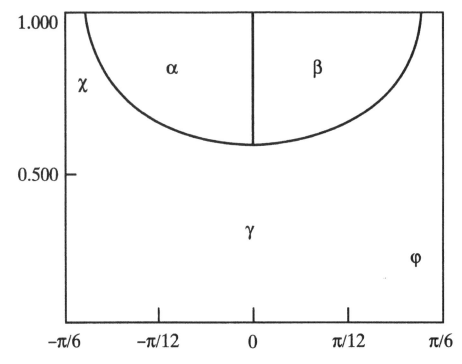

Figure 5.2 Optimal PWM switching law

Table 5.4 Logic table to rename VSI output voltage vectors

θ	φ	U_1	U_2	U_0^-	U_0^+
$0 \cdots \pi/3$	$\theta - \pi/6$	U_1	U_2	U_0^-	U_0^+
$\pi/3 \cdots 2\pi/3$	$\theta - \pi/2$	U_2	U_3	U_0^+	U_0^-
$2\pi/3 \cdots \pi$	$\theta - 5\pi/6$	U_3	U_4	U_0^-	U_0^+
$\pi \cdots 4\pi/3$	$\theta - 7\pi/6$	U_4	U_5	U_0^+	U_0^-
$4\pi/3 \cdots 5\pi/3$	$\theta - 3\pi/2$	U_5	U_6	U_0^-	U_0^+
$5\pi/3 \cdots 2\pi$	$\theta - 11\pi/6$	U_6	U_1	U_0^+	U_0^-

In order to prevent the appearance of frequency beats in the output voltage due to differences in the VSI modulation period, the period of updating data for setting PWM algorithms should be chosen as a multiple of the modulation period of each of the three switching laws (e.g., it can be chosen as three modulation periods of $\alpha(\beta)$ switching law, that corresponds to two modulation periods of the γ switching law).

5.3   OPTIMAL SWITCHING LOSSES IN REAL SLIDING MODE

The challenge of increasing the technical and economic indicators of a "VSI – synchronous motor" in sliding mode can be solved by regularization of the real sliding motion, and it provides some desired properties, such as minimization of switching losses, when using PWM of the first kind (Ryvkin & Isozimov, 1997).

As shown in section 5.2, the VSI control providing switching losses minimization is based on the use of one of three switching laws depending on value and angular position of the VSI reference voltage vector. Each of these laws corresponds to its current error vector $\Delta I(t)$ time plot on the modulation period (figure 5.1).

It is obvious that if at the real sliding mode on the switching manifolds, the limit cycle coincided with the corresponding current error vector $\Delta I(t)$ time plot providing switching losses minimization, the real sliding movement would be regularized, switching frequency of power switches would be a constant, and switching losses minimization would be achieved.

The first necessary condition for the realization of such approach is the presence of information of the reference voltage vector $U_{ref}(t)$. This information can be easily received if we take into consideration the fact that, owing to the load inductive character inherent synchronous motor, the current error is an integrated estimation of a deviation of average VSI output voltage from the reference one:

$$\Delta I(t) = \frac{1}{L} \int_0^T \Delta U(t)dt + C \tag{5.17}$$

where L is the load inductance and C is a constant of integration defined from a condition of an zero current error on the integration period. Making it zero is equivalent to maintaining average output voltage equal to its reference value.

By further analysis for definiteness it is assumed that the reference VSI output voltage vector $U_{ref}(t)$ is located at the bottom part of the triangle formed by vectors U_0, U_1 and U_2, and its relative amplitude χ is more than 0.6. In this case, according to figure 5.2, the minimum switching losses is provided by β switching law with the current error vector $\Delta I(t)$ time plot presented in figure 5.1.

For the design of a limit cycle, we use the vector method. This method essentially consists in partitioning the space of the current error vector $\Delta I(t)$ into areas. Each area is associated with one or more control vectors, which provide movement to the limit cycle and the movement within it. If, as in the PWM law design, there are only three vectors of the output voltage, one of which is zero, the other two are the nearest to a specified output voltage vector, then there will be only three of the current error control vectors

$$\Delta U_1 = U_1 - U_{ref}, \quad \Delta U_2 = U_2 - U_{ref}, \quad \Delta U_0 = -U_{ref} \tag{5.18}$$

presented in figure 5.3.

The description of current errors $\Delta I(t)$ control vectors took advantage of the orthogonal coordinate system offered in section 5.2 (p, r) with an axis p rigidly connected with a VSI reference output voltage vector $U_{ref}(t)$. Amplitudes and angular positions of current error control vectors (5.18) are calculated by using information of the VSI reference voltage vector and the VSI instant ones.

$$d\Delta I/dt = \frac{1}{L}\Delta U_i \tag{5.19}$$

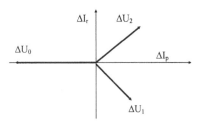

Figure 5.3 Set of vectors of the current error control

The current error space in orthogonal system of coordinates (p, r) is broken by means of nine switching lines in 30 areas. These nine lines belong to three families, each one of them having a central line S_0, S_1 and S_2, passing through the origin of coordinates, and two ones being symmetric, designated by corresponding top indexes $(+)$ and $(-)$:

$$S_0 = \Delta I_r = 0,$$
$$S_0^+ = \Delta I_r - a = 0,$$
$$S_0^- = \Delta I_r + a = 0,$$
$$S_1 = \Delta I_r + k_1 \Delta I_p = 0,$$
$$S_1^+ = \Delta I_r + k_1(\Delta I_p - b) = 0, \quad\quad (5.20)$$
$$S_1^- = \Delta I_r + k_1(\Delta I_p + b) = 0,$$
$$S_2 = \Delta I_r - k_2 \Delta I_p = 0,$$
$$S_2^+ = \Delta I_r - k_2(\Delta I_p + b) = 0,$$
$$S_2^- = \Delta I_r - k_2(\Delta I_r - b) = 0$$

where k_1 and k_2 are angular factors, and a and b are constants:

$$k_1 = \frac{\sin(\pi/6 - \phi)}{\cos(\pi/6 - \phi) - (\sqrt{3}/2)\chi} \quad\quad (5.21)$$

$$k_2 = \frac{\sin(\pi/6 - \varphi)}{\cos(\pi/6 - \varphi) - (\sqrt{3}/2)\chi} \qu\quad (5.22)$$

$$a = 3U_{in}T\sin(\varphi + \pi/6) \times \sqrt{1 - \chi\cos(\pi/6 - \varphi)\sqrt{3} + 3\chi^2/4/4} \qu\quad (5.23)$$

$$b = (1 - \chi\cos\varphi)TA/2 \qu\quad (5.24)$$

By choosing the splitting of the current error space and corresponding control selecting the limit cycle coinciding with current error vector $\Delta I(t)$, the time plot corresponding to a minimum switching loss PWM (figure 5.1, β) can be realized in closed loop, i.e. the real sliding motion with minimum switching losses is achieved.

Table 5.5 Switching loss minimizing sliding mode control

Area Num	Area	Control	Prehistory
1	$S_1 < 0$ & $S_2^+ \geq 0$	U_2	Independent
2	$S_1 \geq 0$ & $S_0^+ \geq 0$	U_1	Independent
3	$S_1 > 0$ & $S_2^+ < 0$ & $S_0 > 0$	U_0	U_2
		U_1	U_1
4	$S_0 < 0$ & $S_0^- > 0$ & $S_2^- > 0$	U_2	U_1
		U_2	U_2
5	$S_1^+ < 0$ & $S_2^- \leq 0$	U_2	Independent
6	$S_2 < 0$ & $S_2^- < 0$ & $S_1 \leq 0$	U_1	Independent
7	$S_2^+ < 0$ & $S_2 \geq 0$ & $S_0 \leq 0$	U_0	Independent
8	$S_0 > 0$ & $S_1 < 0$ & $S_2^+ < 0$	U_0	U_0
		U_1	U_1
		U_2	U_2
9	$S_0 < 0$ & $S_1 > 0$ & $S_2^- > 0$	U_0	U_0
		U_1	U_1
		U_2	U_2

According to the chosen approach in the current error space for generating the desired limit cycle, the areas with both unambiguous and ambiguous sections must be control functions. If an error is in a current error area of the first kind, i.e. it is connected only with one control, this control is always included. At the same time, if a current error is in the current error area of the second type, the included control is determined not only by the area, but by the previous control, i.e., by the control history. A control implementing such an approach is presented in logical table 5.5.

Figure 5.4 shows the partitioning of the current error space on areas of corresponding controls, according to logic table 5.5. An implemented limit cycle is presented.

From the analysis of system trajectories in the current errors space, it is obvious that in the specified above multidimensional relay system there is a real sliding movement on crossing of the chosen switching surfaces with previous reference frequency and the limit cycle corresponding to an operating mode with the minimum switching losses.

The received result, also as well as in case of feedforward PWM, thanks to symmetry of an arrangement of VSI instant output voltage vectors, easily extends on a case of any location of a VSI reference voltage vector $U_{ref}(t)$ by use of logic table 5.4 of renaming of VSI instant voltage vectors.

It is obvious that in the frame of the suggested splitting of the current error space, there is variety of areas that even in case of the unequivocal characteristic connected with none, but some controls that provide a desirable limit cycle. Therefore some switching losses minimizing controls could be suggested. This freedom of choice could be used for the solving additional control problems, such as:

– Switching minimization by the crossing sliding surfaces movement;
– Simplification of control implementation, etc.

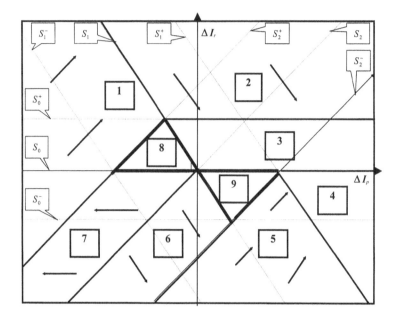

Figure 5.4 Control area and limit cycle

5.4 SWITCHING REGULARIZATION OF DISCONTINUOUS CONTROL VECTOR COMPONENTS

5.4.1 Control vector design

From the viewpoint of abating noise, the VSI switches switching regularity consisting in cyclic transitions between the nearest vectors of instant output voltage VSI (figure 1.5) is quite attractive. Such law leads to serial switching of VSI phase switches (figure 5.5), and the switching sequence remains invariable by all values of an average (during modulation period) voltage vector.

The control problem design consists in working out control of such commands p_j $(j = R, S, T)$ for the VSI phase switches that provide the regulation and desired character (in agreement with figure 5.5) of the VSI instant output voltage vectors switching in real sliding movement.

When implementing a particular vector of instantaneous voltage, a change of direction of the current (or, in the quasi-static mode, at a relatively slow paced setting the current direction of change in the current implementation errors), as shown in section 5.3, determined by the difference vectors of instantaneous voltage and average voltage vector. The selecting criterion for the boundaries of switch control is desirable process of occurrence of real sliding mode, i.e. the process of establishing the sequence and the modulation period for arbitrary initial values of the current error vector.

That dynamic process of establishing the limit cycle characterizing the real sliding mode occupies a finite time and comes to an end for the final number of switching

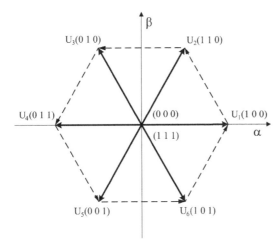

Figure 5.5 Transitions between voltage vectors using nearest vectors of instantaneous voltage (horizontal bars) and average

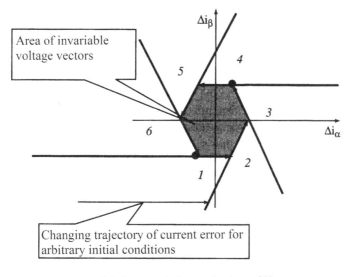

Figure 5.6 Limit cycle for small values of U_{eq}

power switches. It is necessary that area borders coincide with the trajectories realized in real sliding mode.

The power switching function of each VSI phase is characterized by symmetric "hysteresis" by small values of the VSI output voltage vector (figure 5.6).

In contrast to the relay phase current regulator in the vector control, the input of the hysteresis element of each phase receives no error in the phase current, but a linear combination of the phase currents errors.

In a central current error area (neighborhood of the zero current error point) due to hysteresis remain the former values of instant voltage vector control (the control command combinations (000) and (111), corresponding to VSI output zero voltage are forbidden). This form of a limiting cycle in the current error space differed from a correct hexagon by big values of a VSI output average voltage vector (near the maximal value).

As concluded from figure 5.6, the essence of our approach to the design of the vector relay servo system consists in the formation of control switching areas borders in the current error space. These border directions coincide with ones of the difference between the VSI instant voltage vector and the average one, and they are apart by an equal distance from the origin of coordinates. (This distance is the parameter defining the switching frequency). Always using such approach (if only the average voltage vector does not overstep the bounds of the realized voltage vector area), obviously the realization duration of any instant voltage vector in a cycle is not equal to zero. Thus, the same sequence commuting always remains. In this case the duration of the realization of any instant voltage vector in a cycle is not equal to zero. Thus, the same sequence commuting always remains. This is true if the average voltage vector does not go beyond the bounds of the realized voltage vector area.

The offered approach allows designing VSI power switch control for the current closed loop, which is presented below.

Let us define directing vectors of area borders:

$$E_1 = \sqrt{2/3}\, U_{in}(0, -1)^T - (U_{eq\beta}, -U_{eq\alpha})^T,$$

$$E_2 = \sqrt{2/3}\, U_{in}(\sqrt{3}/2, -1/2)^T - (U_{eq\beta}, -U_{eq\alpha})^T,$$

$$E_3 = \sqrt{2/3}\, U_{in}(\sqrt{3}/2, 1/2)^T - (U_{eq\beta}, -U_{eq\alpha})^T,$$

$$E_4 = \sqrt{2/3}\, U_{in}(0, 1)^T - (U_{eq\beta}, -U_{eq\alpha})^T,$$

$$E_5 = \sqrt{2/3}\, U_{in}(-\sqrt{3}/2, 1/2)^T - (U_{eq\beta}, -U_{eq\alpha})^T,$$

$$E_6 = \sqrt{2/3}\, U_{in}(-\sqrt{3}/2, -1/2)^T - (U_{eq\beta}, -U_{eq\alpha})^T \qquad (5.25)$$

It is accepted that the characterizing switching surface by a directing ort e_i $(i = 1, \ldots, 6)$ is orthogonal to the surface (line). To do this, normalize the previous expressions (5.25) by dividing them by the absolute value of the orthogonal switching surface vector, $e_i = E_i / \mathrm{mod}\,(E_i)$. Each surface is separated from the origin in the current error space at an equal distance, denoted by the distance symbol δ. See figure 5.7.

The switching function s_i is formed as the inner product of the current error vector $\Delta i = (\Delta i_\alpha, \Delta i_\beta)$ and the directing ort e_i (i.e. the Δi projection and the directing ort), with the additive component δ, characterizing the distance between the switching surface and the origin:

$$s_i = (\Delta i, e_i) - \delta = \Delta i_\alpha e_{i\alpha} + \Delta i_\beta e_{i\beta} - \delta \qquad (5.26)$$

To obtain the δ estimation let us consider a switching cycle for small values of the average voltage vector ($U_{eq} = 0$). The cycle consists of six motions, the distance moved by the current error vector for each motion is $2\delta/\sqrt{3}$, and the speed of the current error vector is equal to $\sqrt{2/3}U_0$.

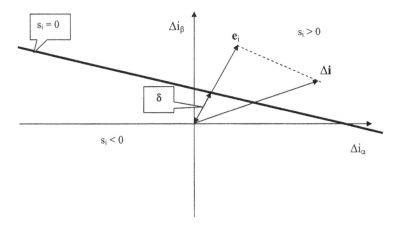

Figure 5.7 Formation of the switching surface

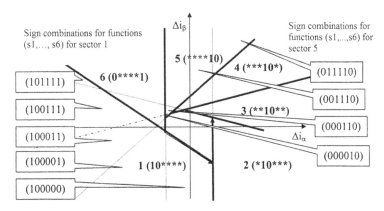

Figure 5.8 Signs combinations of control switching functions

Therefore, the cycle duration T is $12\delta/\sqrt{2}U_0$. Hence, by a reference value of T, the corresponding value δ is determined by

$$\delta = TU_0\sqrt{2}/12 \tag{5.27}$$

The base method of the control switching area borders formation in real sliding movement is described above. We assume now that the switching functions $sg(s_i)$ are generated. According to relay control the signs of these functions (logic signals) in the area must be defined. Required space splitting is defined by crossing the control area borders. The desired partition in sectors is given by the intersection of two neighboring borders of switching control, and the sign of the logical function of one of them is equal to "1" and the other to "0". Possible combinations of logic (sign) functions $(sg(s1), \ldots, sg(s6))$ for these sectors and sector numbers are shown in figure 5.8 (the values of sign functions that do not influence on the area selection are marked with an asterisk ($*$)).

E.g., the current error vector is in the sector 5. As shown the figure 5.8 the signs of the switching functions s_4 and s_5, i.e. $sg(s_4) = 0$, $sg(s_5) = 1$, define clearly the vector position in this sector, then the signs of the other switching functions in this sector are arbitrary (may be equal to 0 and 1, depending on the position of the affix in a particular area in this sector). In sector 1, $sg(s_1) = 1$, $sg(s_2) = 0$, and the signs of the other four switching functions can be equal to "1" or "0". These examples show that in general, to determine the sector in which the current control error vector is, it is necessary to use the values of two (out of six) switching functions. The combination of characters is uniquely determined by the sector of the current error vector.

So, if the affix is in one of six sectors out of the area near the origin of coordinates possible combinations of switching functions signs are characterized by the sequence of the values "1" going successively (with cyclic shift), and "0", also going successively. The position of last "1" in sequence defines the sector number. However, such control structure is fair only by "the correct" sector organization, when crossing their borders defines the convex hexagon that is a limit cycle trajectory of current error changing. In the presence of uncorrelated noise in switching function signals, the "correct" control structure of the switching functions can be broken, and an unequivocal definition of sector on the base of two corresponding switching function signs is impossible. In such cases it is necessary to work out the special measures preventing "false" appointments of VSI switch controls.

Control selection should satisfy the following conditions:

–   It should provide the desired limit cycle, i.e. the sequential switching of VSI power switches phases.
–   A hitting of the image point on the limit switching cycle should be provided from random initial conditions, as well as any changes of the current value of the current error (accidental, caused by noise measurements, or an implementation error, or caused by changes of the current value of the reference).
–   The hitting should occur within a finite time and finite number of power switchings.

The important condition is also noise immunity of this follow-up control plant: presence of noise in input signals of the comparators defining of switching function signs should not lead to changes of VSI switch control commands, at least by small enough noise level. Therefore control selection cannot be unequivocal (on all phases) by an accessory of a current error vector to this or that sector. The presence of inevitable discrepancies and noise by the current measurement could lead to instant (with noise frequency) switch control command changing. There is a necessity to use additional "hysteresis" in the control loop for "cutting" noise of measurement noises with small amplitudes.

Using "hysteresis" assumes the presence of VSI control command memory or switching function memory and implementation of additional conditions of control command selecting similar to switching device hysteresis forming by the scalar relay control. It is convenient to realize hysteresis in vector follow-up control plant by using VSI control command memory. The control shown in figure 5.9 satisfies these conditions.

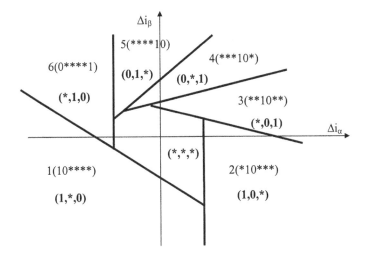

Figure 5.9 Phase switching control command

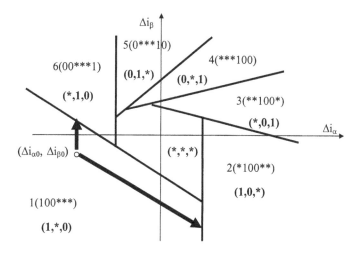

Figure 5.10 Initial movement in sector 1 with arbitrary initial conditions

In figure 5.9 use of boldface indicates VSI (R, S, T) switch control command phases: 1 – to switch-on the line (+), 0 – to switch-on line (–) of a direct current link. The symbol (∗) means that the control command for this phase switch in this sector is not defined. It could be 0 or 1, e.g. previous value stored. It is not defined in the central sector (retain their previous values). The control of all phases, however, control command combinations (0,0,0) or (1,1,1) are forbidden. It avoids the frequent change of phase control commands by the current error affix movement in the neighborhood of any switching control boundary ("vector" hysteresis).

In figure 5.10 for the sector 1 the possible motion trajectories by the initial conditions $(\Delta i_{\alpha 0}, \Delta i_{\beta 0})$ are shown. There are two possible movements depending on the phase S control command (due to symmetry, the movements in other sectors have the same features).

5.4.2 Simplified control

Switching functions (5.26) previously identified as scalar products of a current error vector and a directing ort of the corresponding area border. However, for control design only logic (sign) values of this switching functions $sg(s_i)$ are used. Obviously, the function sign s_i will not change, if it is multiplied by a positive number equal to $mod(E_i)$. Thus, the switching function assumes the following form:

$$s_i = (\Delta i_\alpha e_{i\alpha} + \Delta i_\beta e_{i\beta}) \, mod(E_i) - \delta \, mod(E_i)$$
$$= (\Delta i_\alpha E_{i\alpha} + \Delta i_\beta E_{i\beta}) - \delta \, mod(E_i) \tag{5.28}$$

The first term in the equation (5.28) after substitution of variable E_i, could be rewritten in the form of the sum of scalar product of a current error vector and the directing vectors defining directions of VSI output instant voltage vectors and scalar product of a current error vector and average voltage vector:

$$\Delta i_\alpha E_{i\alpha} + \Delta i_\beta E_{i\beta} = \Delta i_\alpha U_{i\beta} - \Delta i_\beta U_{i\alpha} - \Delta i_\alpha U_{eq\beta} + \Delta i_\beta U_{eq\alpha}$$
$$= \sqrt{2/3} U_0 (\Delta i_\alpha e_{i\beta} - \Delta i_\beta e_{i\alpha}) - \Delta i_\alpha U_{eq\beta} + \Delta i_\beta U_{eq\alpha} \tag{5.29}$$

Since the components of direction instant voltage vectors e_i are predetermined and known, the coefficients before the current error components of the current error Δi_α, Δi_β in the first term can be calculated in advance. For this term calculation adders of current error components can be used. Scalar product of the error current vector and the average voltage vector can be calculated by using adders and two multiplying D/A converters (their digital inputs get slowly changing components of average voltage; their analog inputs get current components). It is very important that the same scalar product be used in calculations of all six of switching functions s_i.

For the calculation of switching functions is also necessary to compute (in the processor) the values of distances between the switching surfaces and the origin of coordinate, i.e. the module of the difference between the instant voltage vectors and the average one). It is useful to make a valuation of switching functions by completing the multiplication by $\sqrt{2/3} U_0$ in the processor part to simplify the hardware.

5.4.3 Follow-up current vector control structure

Hardware implementation of vector relay control included a built-in energy-efficient VSI control. Vector control was has not been widely used mainly due to the complexity of setting up this control, but there still remains an urgent design problem of an electrical drive with digital, software-implemented system of relay-vector control. Such drive would have robustness to the changing semiconductor power converter parameters, with the existing physical constraints of speed, as well as the known advantages

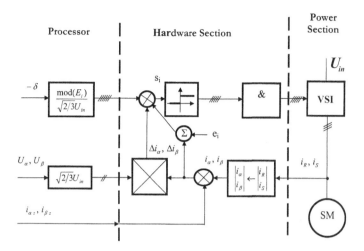

Figure 5.11 Follow-up system structure

of digital control systems, namely: self-testing, auto-tuning, wide front-end and other features.

The structure of the proposed servo drive is shown in figure 5.11. The system contains the following components: current error coordinate transformer, block computation of the switching voltages boundaries, comparators, logical transformers, taking into account the specified switching control. Coordinate transformers can be implemented using D/A converters, so that the current error transformation coefficients could be computed by the control processor.

5.4.4 Test simulation of a follow-up loop

A test simulation of the proposed follow-up control was carried out with the following goals:

– Verify the follow-up loop is functioning;
– Identify operational mode areas;
– Define follow-up errors and the design of their compensation methods;
– Identify the sensitivity of the PWM follow-up control to inaccuracies formation of switching functions;
– Determine the extent of follow-up loop noise immunity;
– Requirements definition to the hardware and software PWM blocks (speed, capacity, word length, etc.);
– Identify possibilities to simplify the hardware;
– Develop proposals for implementing the follow-up PWM control.

Regarding the use of a follow-up PWM control in electrical drives, it should be noted that the follow-up PWM, or the VSI with a closed loop PWM that could be named controlled current source, is aimed for use in fairly complex drives with vector

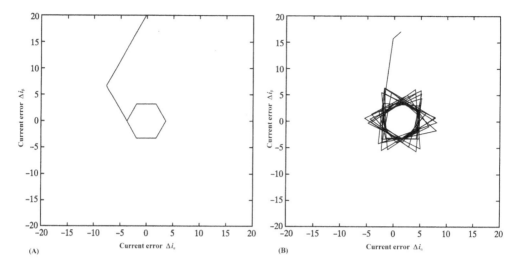

Figure 5.12   Current error vector time plots by various values of the VSI average output voltage. A) zero, B) 80% from the maximum realizable value per unit

control. Electrical drive structure is not important by the PWM control simulation. So it makes sense to conduct simulations for a simplified scheme of the power unit, reflecting substantial (for the control design of the closed loop PWM) features of the VSI load. Such simplification may be in the representation of VSI three-phase load as two-dimensional one (in the generalized system) and an inductive load connected in series with it, and a voltage source with two independent components. The values of the latter are exactly the values of the components of the equivalent voltage.

This load is connected in series with a voltage source with two independent components. Their values are exactly the values of the components of the equivalent voltage. In real electrical drives, the values of fundamental frequency of the VSI output voltages and of the PWM frequency are selected (tens to hundreds of Hz for the first and the units-tens of kHz for the second). It allows by simulation to use a constant (quasi-statically varying) voltage source in the VSI load, as well as constant (quasi-statically changing) current component references for the closed loop PWM control. These assumptions allow closing the PWM loop without taking into account the specifically features of the motor being used.

A simulation was done using Matlab/Simulink and the results are shown in figure 5.12.

The presented current error vector time plots show that the suggested switch control provides regular structure of VSI phase serial switching and it is not dependent on the average value, (during switching) value, of the VSI output voltage vector. Also, by the small values of a VSI output average voltage, thanks to symmetric "hysteresis" of the switching functions of each VSI phase power switch a limit cycle of self-oscillations, i.e. a correct hexagon has been established at once. In contraposition to it, switching function (5.26) has non-symmetric hysteresis on large VSI output average voltage values. The process of establishing a limit cycle is extended, and the limit cycle differs a little from a correct hexagon.

REFERENCES

Akkaya, R., Yildirmaz, G. and Gulgun, R. "*A space vector modulation technique with minimum switching loss for VSI PWM inverters*". Proc. of the 7th International Power Electronics & Motion Control Conference, PEMC'96, Budapest, Hungary, 1996, vol. 2, pp. 352–355.

Holz, J. "*Pulsewidth modulation for electronic power conversion*". Proc. of the IEEE, 1994, vol. 82, no. 8, pp. 1194–1213.

Kuntsevich, V.M. and Chehovoy, Ju.N. "*Nonlinear control with frequency-pulse-width modulation*". Kiev: Technika, 1970. 340 p. (in Russian).

Mohan, N., Underland, T.M. and Robbins, W.P. "*Power electronics: converters, applications and design*". 3rd ed. New York: John Wiley & Son Inc., 2003. 824 p.

Nagy, I. "*Improved current controller for PWM inverter drives with the background of chaotic dynamics*". Proc. the 20th International Conference on Industrial Electronics Control and Instrumentation, IECOM'94, Bologna, Italy, 1994, pp. 561–566.

Pfaff, G. and Wick, A. "*Direkte Stromregelung bei Drehstromantrieben mit Pulswechselrichter*". Regelungstechnische Praxis, 1983, Bd. 4, N. 11, S. 472–477.

Ryvkin, S. and Izosimov, D. "*Comparison of pulse-width modulation algorithms for three-phase voltage inverters*". Electrical Technology, 1997, no. 2, pp. 133–144.

Ryvkin, S. and Izosimov, D. "*Novell switching losses optimal sliding mode control technique for three-phase voltage source inverter*". Proc. IEEE International Symposium on Industrial Electronics, ISIE'97, Guimaraes, Portugal, 1997, vol. 2, pp. 288–293.

Ryvkin, S.E., Belkin, S.V. and Izosimov, D.B. "*Commutation laws transfer strategy for the feedforward switching losses optimal PWM for three-phases voltage source inverter*". Proc. 24th Annual Conference of IEEE Industrial Electronics Society, IECON'98, Aachen, Germany, 1998, pp. 768–772.

Shevtsov, S.V., Izosimov, D.B. and Ryvkin, S.E. "*Space-vector simplex pulse-width modulation methods of 3-phase voltage source inverter control*". Proceedings of the 20th International Conference on Industrial Electronics, Control and Instrumentation IECON'94, Bologna, Italy, 1994, pp. 520–525.

Suetz, Z., Nagy, I., Backhauz, L. and Zaban, K. "*Controlling chaos in current forced induction motor*". Proc. the 7th International Power Electronics & Motion Control Conference, PEMC'96, Budapest, Hungary, 1996, vol. 3, pp. 282–286.

Trzynzdlowski, S.V. and Legowski, S. "*Minimum-loss vector PWM strategy for three-phase inverters*". IEEE Transactions on Power Electronics, 1994, vol. 9, no. 1, pp. 26–34.

Zinoviev, G.S. "*Power electronics backgrounds*". Novosibirsk: Publishing house of Novosibirsk state technical university, 2005. 664 p. (in Russian).

Chapter 6

Mechanical coordinates observers

6.1 GENERAL FORMULATION OF THE OBSERVATION PROBLEM

Obtaining information about the state of the control plant, particularly about the control variables, is of key importance in control design. It is especially important when designing a high quality, high precision electrical drive, requiring both static and dynamic limits. The traditional approach based on direct measurement of all necessary control coordinates leads to considerable complication of the electrical drive design, which deteriorates its operation and cost indexes. It is possible to overcome these limitations by excluding sensors for those variables whose direct measurement is undesirable. So, they are estimated using observers instead, and used in control design (Consoli, 2000), (Dote, 1988), (Luenberger, 1966), (Holtz, 2004).

Linear observers of mechanical coordinates using the results of the linear theory of observation found wide acceptance. Such an observer is a dynamic system designed to use the dynamic model of the mechanical part of an electrical drive (1.1). Its inputs comprise values of the electromagnetic motor torque and some of the measured mechanical coordinates.

The next step is the full exclusion of mechanical coordinate sensors and the construction of the electrical drive to contain only electric variables sensors. The complexity of such an approach lies in the electromagnetic part of an electric motor, in particular the synchronous motor, as shown in section 1.1.1, as described by the nonlinear equations (1.4)–(1.8). To obtain estimates of mechanical variables using only electrical ones, new nonlinear methods for estimating mechanical coordinates must be developed.

For this estimation problem solving the results of the theory of systems with sliding modes (Dote, 1988), (Utkin, 1992), (Utkin et al, 1999), (Andreescu et al, 2000), (Vittek & Dodds, 2003) could be maintained in perspective. As shown in chapter 3, this approach allows effective decomposition of an observation problem by simplifying an information processing system that is rather essential, taking into account the nonlinearity and complexity of the electrical drive (high order of equations). It must be emphasized that the main idea of sliding mode observer is designing a special dynamics system with the discontinuous controls, in which the sliding mode is willfully organized. The equivalent controls, i.e. an average values of the discontinuous ones, should give the information about the mechanical coordinates. However, the values of the equivalent controls could be received by using a filter with its own dynamics. Therefore, it must be taken into account that the estimation of a mechanical coordinates is possible in case that the filter is singular perturbed in relation to mechanical subsystem (1.1), i.e. the rates of a filtration should be essentially above (by an order of magnitude) the rates of change of the estimated mechanical coordinates.

The observer giving estimations of unmeasured variables usually represents an imitating model of the investigated plant. The approximation degree of the model of the plant is defined, first of all, by the available information about the plant variables. It is obvious that the model structure and algorithms to receive estimates of mechanical coordinates, especially in the case of a nonlinear plant such as a synchronous motor, essentially depend on motor type. Therefore, we will consider the observer design for two types of synchronous motors, which will be considered separately (Ryvkin & Isozimov, 1994), (Ryvkin, 1995), (Ryvkin, 1996), namely:

–   Permanent magnet *nonsalient-pole* synchronous motor;
–   Synchronous reluctance motor.

Observation design algorithms will be performed, both in the rotating (d, q), and in the motionless (α, β) system of coordinates. As a known reference information, the information about a stator current vector I and a stator voltage vector U (or, equivalently, the actual phase currents and voltages), whose measurement is not difficult, and the parameters of electrical circuits. It is necessary to receive estimations of the angular rotor position Γ and the angular speed Ω coordinates, which are proportional to the angle γ_R between the motionless axis R of three-phase fixed coordinate system and the mobile d axis of the rotating coordinate system and the electrical angular speed of the rotor ω, respectively ($\gamma_R = p\Gamma$; $\omega = p\Omega$, where p is a number of pair poles of the motor). For simplicity we shall assume henceforth that a synchronous motor has two poles, i.e. the electric and mechanical coordinates are equal.

6.2 OBSERVER DESIGN FOR PERMANENT MAGNET SALIENT-POLE SYNCHRONOUS MOTOR WITH CONSTANT MAGNETS

6.2.1 Rotating coordinate system

The electromagnetic part of a permanent magnet *nonsalient-pole* synchronous motor in the rotating coordinate system (d, q), connected to its rotor, according to (1.8), is described by the following system of differential equations:

$$\frac{di_d}{dt} = -\frac{ri_d}{L} + \Omega i_q + \frac{u_d}{L}$$

$$\frac{di_q}{dt} = -\frac{ri_q}{L} - \Omega \cdot \left(\frac{\psi}{L} + i_d\right) + \frac{u_q}{L} \qquad (6.1)$$

$$\frac{d\Gamma}{dt} = \Omega$$

The components of the current and voltage vectors in the rotating coordinate system are related to the components of the current and voltage vectors in a motionless coordinate system by a transformation (1.2), (1.3). The information about a current value of an angular rotor position Γ is used in it. Because the value of the angular rotor position Γ is not known in advance but depends on the values of the current and voltage in the rotating coordinate system, we will take advantage of its estimation $\hat{\Gamma}$, assuming that it exists. Furthermore, we shall distinguish actual variables from their corresponding estimates using the caret symbol "^" above the name of the estimated variable.

The interesting electrical variables are in this case the current and the voltage in the rotating coordinate system, since they are not calculated using the actual rotor angular position Γ anymore, but its estimation $\hat{\Gamma}$. They are described as follows:

$$
\begin{aligned}
u_d^* &= u_\alpha \cos \hat{\Gamma} + u_\beta \sin \hat{\Gamma}, \\
u_q^* &= u_\alpha \sin \hat{\Gamma} + u_\beta \cos \hat{\Gamma}, \\
i_d^* &= i_\alpha \cos \hat{\Gamma} + i_\beta \sin \hat{\Gamma}, \\
i_q^* &= i_\alpha \sin \hat{\Gamma} + i_\beta \cos \hat{\Gamma}
\end{aligned}
\tag{6.2}
$$

where variables with a slash "*" above represent voltage and current vectors components in the rotating coordinate system. They are calculated from the appropriate measured components of the current and voltage vectors in the motionless coordinate system using the estimation value of the angular rotor position $\hat{\Gamma}$.

Let us assume that the value of the estimated angular rotor position is closely approximated to the actual one, i.e. an angular rotor position error $\Delta\Gamma = \Gamma - \hat{\Gamma}$ is insignificant. In this case, taking into account (6.2), the synchronous motor behavior is described, not by equations (6.1), but by the following equations:

$$
\begin{aligned}
\frac{di_d^*}{dt} &= -\frac{ri_d^*}{L} + \frac{u_d^*}{L} + i_q^* \left(\frac{d\hat{\Gamma}}{dt}\right) + \frac{\psi\Omega\Delta\Gamma}{L}, \\
\frac{di_q^*}{dt} &= -\frac{ri_q^*}{L} - \frac{\Omega\psi}{L} + \frac{u_q^*}{L} - i_d^* \left(\frac{d\hat{\Gamma}}{dt}\right)
\end{aligned}
\tag{6.3}
$$

As mentioned above in section 6.1, the mechanical variables information should be received using the information on the measured stator currents and voltages. In this case, the imitating model must be constructed as close as possible to the observable control plant structure. The base for such a model is the mathematical description of the synchronous motor (6.3). The calculated voltage components values are used as input information. In order to make the motor model behavior almost the same as the real one, the values of the model current components must be equal to the calculated values of the motor current components. Two correcting controls u_1 and u_2 on the number of model variables could be used. The following emulation model comes up as a result:

$$
\begin{aligned}
\frac{d\hat{i}_d}{dt} &= -\frac{r\hat{i}_d}{L} + \frac{u_d^*}{L} + i_q^* \left(\frac{d\hat{\Gamma}}{dt}\right) + u_1 \\
\frac{d\hat{i}_q}{dt} &= -\frac{r\hat{i}_q}{L} + \frac{u_q^*}{L} - i_d^* \left(\frac{d\hat{\Gamma}}{dt}\right) + u_2
\end{aligned}
\tag{6.4}
$$

The suggested model (6.4) will correctly reflect the behavior of the investigated control plant, (6.3), if the behavior of the model current stator components $\hat{i}_d$, $\hat{i}_q$ and the real system i_d^*, i_q^* repectively coincide. It is possible to achieve this using correcting controls u_1 and u_2 that are discontinuous and must guarantee the sliding mode on the intersection of the sliding motion surfaces, which present zero errors for the appropriate components of the stator current:

$$S_d = i_d^* - \hat{i}_d = 0$$
$$S_q = i_q^* - \hat{i}_q = 0$$

(6.5)

In this case, matrix A, before the vector of correcting controls in equation (4.2) projecting initial motion of dynamic systems (6.3) and (6.4) on the subspace of zero errors (6.5), is diagonal. The initial problem is the design of sliding mode on the intersection of the switching surfaces $S_d = 0$ and $S_q = 0$, which breaks up into two independent one-dimensional ones. Each of them deals with the use of existence conditions of one-dimensional sliding movement (2.6). To maintain sliding movement on crossing the above specified sliding surfaces, correcting controls u_1 and u_2 must be discontinuous in nature:

$$u_1 = U_1 \operatorname{sgn} S_d$$
$$u_2 = U_2 \operatorname{sgn} S_q$$

(6.6)

Additionally, their amplitudes U_1 and U_2 are selected from the sliding mode existence condition:

$$U_1 \geq \left| -\frac{r(i_d^* - \hat{i}_d)}{L} + \frac{\Omega \psi \Delta \Gamma}{L} \right|$$

$$U_2 \geq \left| -\frac{r(i_q^* - \hat{i}_q)}{L} + \frac{\Omega \psi}{L} \right|$$

(6.7)

In a sliding mode the errors S_d and S_q are equal to zero, and the correcting controls u_1 and u_2 are changed with infinite frequency. The continuous analogues of these discontinuous controls, which are named equivalent controls, u_{1eq} and u_{2eq}, owing to equations (6.3)–(6.5), are defined as follows:

$$u_{1eq} = \frac{\Omega \psi \Delta \Gamma}{L}$$

(6.8)

$$u_{2eq} = \frac{\psi \Omega}{L}$$

(6.9)

Obviously, the equivalent value of the correcting control u_2 is an estimate of the rotor speed Ω, and the corrective value u_1 is a test signal. Its difference from the zero value can be judged on the inconsistency of the used model or its initial conditions and the actual synchronous motor. The proposed model (6.5) will completely describe the electromagnetic processes in the real synchronous motor (6.4) in the event that an error in determining the angular rotor position is zero ($\Delta\Gamma = 0$) or equivalently, the equivalent value of the correcting control u_1 is zero.

For its maintenance, it is necessary to add used model (6.5) with the model of a mechanical part with an estimation of the initial conditions.

The model of a mechanical part is designed based on the equation of mechanical movement of the rotor connecting the angular rotor positions Γ and angular speed Ω with use of the correcting signal v:

$$\frac{d\hat{\Gamma}}{dt} = \hat{\Omega} + v \tag{6.10}$$

The correcting signal v urged to provide coincidence of estimations of the rotor speed, received with use of the electromagnetic model (6.5) and the model of a mechanical part (6.10).

For the design of the correcting signal v we will take advantage of an available signal u_{1eq} (6.8), linearly depending on an angular rotor position error:

$$v = k u_{1eq} \tag{6.11}$$

Taking into account (6.10), (6.11), the behavior of an angular rotor position error in sliding mode on the crossing of the switching surfaces $S_d = 0$ and $S_q = 0$ is described by the following differential equation:

$$\frac{d\Delta\Gamma}{dt} + \frac{k\Omega\psi\Delta\Gamma}{L} = 0 \tag{6.12}$$

It is obvious that in sliding mode the angular rotor position error $\Delta\Gamma$ will tend to zero if the coefficient before the angular position error $\Delta\Gamma$ is greater than zero. Taking into account that the equivalent value of the correcting control u_{2eq} is an estimation of the angular speed Ω (6.9), the condition specified above will always be carried out, if the coupling coefficient k is selected as follows:

$$k = k_0 \, \mathrm{sgn}(u_{2eq}), \quad k_0 - const, \; k_0 > 0 \tag{6.13}$$

From analysis of the equation (6.12) it is possible to draw a conclusion that at zero angular speed is impossible on measurements of currents and voltage in stator winding to define angular position and angular speed of the rotor. From here, in particular, below speed limit for the mechanical coordinates observer of such synchronous motor follows. Numerical values of an admissible range of angular speed, apparently, are expedient for establishing experimentally as they are defined by many factors, such as inadequacy of the real motor and its mathematical model, an error in definition of electromagnetic circuit parameters etc. Nonetheless, because of high-frequency power switching in modern power converters (for example, when using PWM output voltage in VSI), angular rotor speed equality to zero is excluded. Therefore it is possible to expect estimation of necessary technical coordinates estimations in all operating modes of a drive.

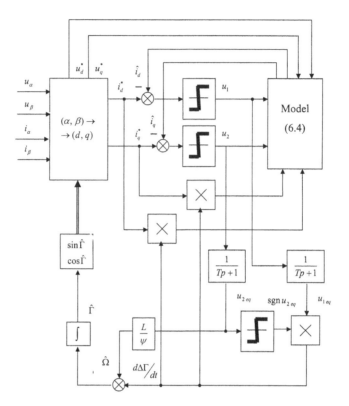

Figure 6.1 Block diagram of the mechanical coordinate observer for the permanent magnet *nonsalient-pole* synchronous motor

The structure of the mechanical coordinate observer for the permanent magnet *nonsalient-pole* synchronous motor is presented in figure 6.1.

6.2.2 Motionless coordinate system (α, β)

The electromagnetic processes in synchronous motors in a motionless coordinate system (α, β) are described by the following system of differential equations

$$
\begin{aligned}
\frac{di_\alpha}{dt} &= -\frac{ri_\alpha}{L} + \frac{\Omega\psi\sin\Gamma}{L} + \frac{u_\alpha}{L} \\
\frac{di_\beta}{dt} &= -\frac{ri_\beta}{L} - \frac{\Omega\psi\cos\Gamma}{L} + \frac{u_\beta}{L}
\end{aligned}
\tag{6.14}
$$

The values of rotor angular position Γ and the rotor speeds Ω in (6.14), as well as in the previous case, are unknown. For their definition, as well as in the previous case, we will design a dynamic system, which must have a structure as close as possible to the structure of the control plant under study (6.14) and in which the available information about the stator voltage and current is used. In order to update the behavior of the model, as well as in the previous case, we will enter two correcting modeling controls u_1 and u_2

$$\frac{d\hat{i}_\alpha}{dt} = -\frac{r\hat{i}_\alpha}{L} + \frac{u_\alpha}{L} + u_1$$

$$\frac{d\hat{i}_\beta}{dt} = -\frac{r\hat{i}_\beta}{L} + \frac{u_\beta}{L} + u_2$$

(6.15)

In order for that model (6.15) to correctly reflect the processes proceeding in the synchronous motor, it is necessary that movements in real (6.14) and modeling (6.15) electromagnetic systems coincide, i.e. the behavior of the model stator current components $\hat{i}_\alpha$, $\hat{i}_\beta$ and i_α, i_β coincide respectively. One possible way to maintain this condition, is ensuring that the sliding mode on the crossing of sliding surfaces represent zero errors on the current components S_α, S_β, similar to (6.10), but in a motionless system of coordinates, by using correcting controls u_1 and u_2

$$S_\alpha = i_\alpha - \hat{i}_\alpha = 0$$

$$S_\beta = i_\beta - \hat{i}_\beta = 0$$

(6.16)

As in the previous case, matrix A is diagonal before a vector of correcting controls in equation (4.2) and projections of movement of initial dynamic systems (6.14) and (6.15) on a subspace of zero errors mismatches (6.16). The initial problem of sliding mode design, on the crossing the switching surfaces $S_\alpha = 0$ and $S_\beta = 0$ breaks up into two independent one-dimensional ones. Each of them is solved using the existence conditions of one-dimensional sliding motion (2.6). To provide a sliding motion over the intersection of the above mentioned sliding surfaces, correcting controls u_1 and u_2 should be discontinuous in nature:

$$u_1 = U_1 \operatorname{sgn} S_\alpha$$

$$u_2 = U_2 \operatorname{sgn} S_\beta$$

$$U_1 \geq \frac{1}{L}|-r(i_\alpha - \hat{i}_\alpha) + \Omega\psi \sin\Gamma|$$

$$U_2 \geq \frac{1}{L}|-r(i_\beta - \hat{i}_\beta) - \Omega\psi \cos\Gamma|$$

(6.17)

The error functions S_α, S_β in sliding mode are equal to zero, and the equivalent values of the controls u_1 and u_2 accept the following values:

$$u_{1eq} = -\Omega\psi \sin\Gamma$$

$$u_{2eq} = \psi\Omega \cos\Gamma$$

(6.18)

They also contain, as well as in the previous case, the information about the rotor angular speed and position.

Unfortunately, in this case, the equivalent values of each of controls contain the multiplication information about the angular position and speed. Therefore it is necessary to carry out additional computing operations to obtain the necessary information about the angular position and speed:

$$\Gamma = -\text{arctg}\left(\frac{u_{1eq}}{u_{2eq}}\right) \tag{6.19}$$

$$\Omega = \frac{\sqrt{u_{1eq}^2 + u_{2eq}^2}}{\psi} \tag{6.20}$$

The received estimations of mechanical variables demand allocation of equivalent values of correcting controls u_1 and u_2 (e.g., filtering low frequencies) and their further processing according to equations (6.19), (6.20).

6.2.3 The simplified observer

It is known that rates of the electromagnetic and electromechanical processes, taking place inside synchronous motors, essentially differ from each other. Therefore, by calculating the values of mechanical variables it is possible to determine the electromagnetic processes established. Under this assumption, the behavior of the synchronous motor is described by the following system of equations:

$$-\frac{ri_d}{L} + \Omega i_q + \frac{u_d}{L} = 0$$

$$-\frac{ri_q}{L} - \Omega \cdot \left(\frac{\psi}{L} + i_d\right) + \frac{u_q}{L} = 0 \tag{6.21}$$

$$\frac{d\Gamma}{dt} = \Omega$$

The estimated variables, assuming a small error of the rotor angular position estimation $\Delta\Gamma = \Gamma - \hat{\Gamma}$ and taking into account (6.2), are related to the measured electric variables according to the following algebraic equations:

$$-ri_d^* + u_d^* + \Omega L i_q^* + \psi\Omega\Delta\Gamma = 0 \tag{6.22}$$

$$-ri_q^* + u_q^* - \Omega L i_d^* - \psi\Omega = 0 \tag{6.23}$$

Using these algebraic connections between the measured electric and the defined mechanical coordinates, it is possible to calculate the value of the angular speed Ω

$$\Omega = \frac{-ri_q^* + u_q^*}{Li_d^* + \psi} \tag{6.24}$$

and one of the rotor angular position errors $\Delta\Gamma$

$$\Delta\Gamma = \frac{ri_d^* - u_d^* - \Omega L i_q^*}{\psi\Omega} \tag{6.25}$$

In order to improve the estimations of angular speed and position, received by means of (6.24) and (6.25), it is possible to use an observer of mechanical movement, similar to (6.10).

The simplified observer is advantageous in digital control. Proceeding from demanded dynamic properties of the electrical drive, the duration of a cycle of control calculation, as a rule, is large enough in comparison to the time constant of the motor electromagnetic circuits. It justifies neglecting the electromagnetic circuit dynamics in this case.

6.3   OBSERVER DESIGN FOR THE SYNCHRONOUS RELUCTANCE MOTOR

6.3.1   Rotating coordinate system

The behavior of an electromagnetic part of this synchronous reluctance motor in a rotating coordinate system (d, q), connected to a rotor, according to (1.6), is described by the following system of the differential equations:

$$\frac{di_d}{dt} = -\frac{ri_d}{L_d} + \frac{L_q}{L_d}\Omega i_q + \frac{u_d}{L_d}$$

$$\frac{di_q}{dt} = -\frac{ri_q}{L_q} + \frac{L_d}{L_q}\Omega i_d + \frac{u_q}{L_q} \qquad (6.26)$$

$$\frac{d\Gamma}{dt} = \Omega$$

Currents and voltages in a rotating coordinate system are related to currents and voltages in a motionless coordinate system through a transformation using the information on the actual rotor angle position (1.2), (1.3). The value of the rotor angular position Γ, also as well as in a case of the permanent magnet *nonsalient-pole* synchronous motor, is not known in advance and must be estimated. For calculation of the values of currents and voltages in a rotating coordinate system (d, q) by using their values in motionless coordinate system, i.e. actually on phase values, we will take advantage of an estimation of the rotor angular position $\hat{\Gamma}$, assuming that it exists. Thus, the electric variables interesting for us are the stator current and voltage vectors. Their components are described according to (6.2) in the rotating coordinate system related not to the actual rotor angle Γ anymore, but its estimation $\hat{\Gamma}$.

Assuming that the error in the definition of the rotor angular position $\Delta\Gamma = \Gamma - \hat{\Gamma}$ is insignificant, the behavior of the electromagnetic part of the synchronous reluctance motor is described by the following equations

$$\frac{di_d^*}{dt} = -\frac{ri_d^*}{L_d} + \frac{u_d^*}{L_d} + i_q^*\left(\frac{d\hat{\Gamma}}{dt}\right) + \frac{(L_d - L_q)}{L_d}\left\{\frac{[-ri_q^* + u_q^* - \Omega i_d^*(L_d + L_q)]\Delta\Gamma}{L_q} - \Omega i_q^*\right\}$$

$$\frac{di_q^*}{dt} = -\frac{ri_q^*}{L_q} + \frac{u_q^*}{L_q} - i_d^*\left(\frac{d\hat{\Gamma}}{dt}\right) + \frac{(L_d - L_q)}{L_q}\left\{\frac{[-ri_d^* + u_d^* - \Omega i_q^*(L_d + L_q)]\Delta\Gamma}{L_d} - \Omega i_d^*\right\}$$

$$(6.27)$$

As well as in the case of the permanent magnet *nonsalient-pole* synchronous motor, to obtain information about the mechanical variables, we will take advantage of a model, whose structure is adequate to the structure of the considered synchronous machine and uses available information about the electric variables of the synchronous motor. In order for the model to correctly reflect the behavior of the investigated plant, it is necessary that the movements in the real (6.27) and the modeling electromagnetic systems coincide, i.e. the behavior corresponding to the model $\hat{i}_d$, $\hat{i}_q$ and the real i_d^*, i_q^* stator current components must coincide.

In order to update the behavior of the model and maintain the coincidence of the stator model and motor currents specified above, a model using the input information about stator voltages will be controlled by two correcting control signals, u_1 and u_2. Taking into account the above mentioned, the model equations assume the following form:

$$
\begin{aligned}
\frac{d\hat{i}_d}{dt} &= -\frac{r\hat{i}_d}{L_d} + \frac{u_d^*}{L_d} + i_q^*\left(\frac{d\hat{\Gamma}}{dt}\right) + \frac{(L_d - L_q)}{L_d}u_1 \\
\frac{d\hat{i}_q}{dt} &= -\frac{r\hat{i}_q}{L_q} + \frac{u_q^*}{L_q} - i_d^*\left(\frac{d\hat{\Gamma}}{dt}\right) + \frac{(L_d - L_q)}{L_q}u_2
\end{aligned}
\tag{6.28}
$$

Coincidence of model behavior with the real plant will be provided, as well as in the previous case, at the expense of the organization of sliding mode on the crossing of the sliding surfaces representing zero errors of the components of stator current in rotating system of coordinates (6.5). The relay controls are the same as (6.6), however, the correcting control amplitudes are:

$$
\begin{aligned}
U_1 &\geq \left| -\frac{r(i_d^* - \hat{i}_d)}{(L_d - L_q)} + \left\{ \frac{[-ri_q^* + u_q^* - \Omega i_d^*(L_d + L_q)]\Delta\Gamma}{L_q} - \Omega i_q^* \right\} \right| \\
U_2 &\geq \left| -\frac{r(i_q^* - \hat{i}_q)}{(L_d - L_q)} + \left\{ \frac{[-ri_d^* + u_d^* - \Omega i_q^*(L_d + L_q)]\Delta\Gamma}{L_d} - \Omega i_d^* \right\} \right|
\end{aligned}
\tag{6.29}
$$

The equivalent value of the correcting control u_{1eq} characterizing sliding mode and one of the correcting controls u_{2eq}, owing to the equations of the plant (6.27), of the model (6.28) and the switching surfaces (6.5) are defined as follows:

$$
\begin{aligned}
u_{1eq} &= \frac{[-ri_q^* + u_q^* - \Omega i_d^*(L_d + L_q)]\Delta\Gamma}{L_q} - \Omega i_q^* \\
u_{2eq} &= \frac{[-ri_d^* + u_d^* - \Omega i_q^*(L_d + L_q)]\Delta\Gamma}{L_d} - \Omega i_d^*
\end{aligned}
\tag{6.30}
$$

Obviously, the equivalent values of correcting controls u_{1eq} and u_{2eq}, as in the previous case, contain information about the rotor angular position and speed. However, in this case, the equivalent value of each control depends on the estimated variables, namely the angular rotor position and the speed. It is therefore necessary to carry out, given the small angular error $\Delta\Gamma$, additional computing operations to obtain the required information on the angular position and speed.

$$\Delta\Gamma = \frac{L_d L_q(i_d^* u_{1eq} - i_q^* u_{2eq})}{L_d i_d^*(-ri_q^* + u_q^*) - L_q i_q^*(-ri_d^* + u_d^*) + (L_d + L_q)(L_d i_d^* u_{2eq} - L_q i_q^* u_{1eq})}$$

$$\Omega = \frac{\begin{array}{c} L_q i_q^*[L_d(-ri_q^* + u_q^*)u_{2eq} - L_q(-ri_d^* + u_d^*)u_{1eq}] \\ +(L_d + L_q)(L_d i_d^* u_{2eq} - L_q i_q^* u_{1eq})L_q u_{1eq} \end{array}}{i_q^*[L_d i_d^*(-ri_q^* + u_q^*) - L_q i_q^*(-ri_d^* + u_d^*)] + (L_d + L_q)L_q[L_d(i_d^*)^2 - L_q(i_q^*)^2]u_{1eq}}$$

$$(6.31)$$

The resulting estimates of mechanical variables requires obtaining the equivalent correcting controls u_{1eq} and u_{2eq} values (e.g. by low-pass filter) and further processing according to expressions (6.31). From a standpoint of high-dynamic properties of control plant and reduction of computation work, it seems that it is more appropriate to take correcting controls μ_1 and μ_2, and the associated linear transformation of the originally selected model and correcting controls u_1 and u_2, i.e.

$$\begin{vmatrix} \mu_1 \\ \mu_2 \end{vmatrix} = \begin{vmatrix} L_d L_q i_d^* & -L_d L_q i_q^* \\ (L_d + L_q)L_q i_q^* & (L_d + L_q)L_d i_d^* \end{vmatrix} \begin{vmatrix} u_1 \\ u_2 \end{vmatrix} \qquad (6.32)$$

In this case the new correcting influences μ_1 and μ_2, are included simultaneously into both equations of the model:

$$\frac{d\hat{i}_d}{dt} = -\frac{ri_d}{L_d} + \frac{u_d^*}{L_d} + i_q^* \left(\frac{d\hat{\Gamma}}{dt}\right) + \frac{(L_d - L_q)}{L_d[L_d(i_d^*)^2 + L_q(i_q^*)^2]} \left(\frac{i_d^*}{L_q}\mu_1 + \frac{i_q^*}{L_d + L_q}\mu_2\right)$$

$$\frac{d\hat{i}_q}{dt} = -\frac{ri_q}{L_q} + \frac{u_q^*}{L_q} - i_d^* \left(\frac{d\hat{\Gamma}}{dt}\right) + \frac{(L_d - L_q)}{L_d[L_d(i_d^*)^2 + L_q(i_q^*)^2]} \left(-\frac{i_q^*}{L_d}\mu_1 + \frac{i_d^*}{L_d + L_q}\mu_2\right)$$

$$(6.33)$$

According to the theorem of invariance of the equations of sliding mode to linear transformations of control (2.12) in the system (6.33) with the new controls the sliding mode on crossing of the selected sliding surfaces (6.5) exists. In this case, the new modeling correcting controls μ_1 and μ_2 represent a combination of initial discontinuous correcting controls u_1 and u_2 (6.29). According to (6.32) their equivalent values, they are equal to:

$$\mu_{1eq} = \{L_d i_d^*(-ri_q^* + u_q^*) - L_q i_q^*(-ri_d^* + u_d^*) - (L_d + L_q)[L_d(i_d^*)^2 + L_q(i_q^*)^2]\Omega\}\Delta\Gamma$$

$$\mu_{2eq} = (L_d + L_q)\{-r[(i_d^*)^2 + (i_q^*)^2] + u_d^* i_d^* + u_q^* i_q^*\}\Delta\Gamma$$

$$- (L_d + L_q)[L_d(i_d^*)^2 + L_q(i_q^*)^2]\Omega \qquad (6.34)$$

Using these equivalent values it is possible to define an error of the rotor angular position $\Delta\Gamma$ and the angular speed Ω:

$$\Delta\Gamma = \frac{\mu_{1eq}}{L_d i_d^*(-r i_q^* + u_q^*) - L_q i_q^*(-r i_d^* + u_d^*) + \mu_{2eq}}$$

$$\Omega = \frac{-\mu_{2eq}[L_d i_d^*(-r i_q^* + u_q^*) - L_q i_q^*(-r i_d^* + u_d^*) + \mu_{2eq}]}{+(L_d + L_q)\{-r[(i_d^*)^2 + (i_q^*)^2] + u_d^* i_d^* + u_q^* i_q^*\}\mu_{1eq}}{(L_d + L_q)[L_d(i_d^*)^2 + L_q(i_q^*)^2][L_d i_d^*(-r i_q^* + u_q^*) - L_q i_q^*(-r i_d^* + u_d^*) + \mu_{2eq}]}$$

$$(6.35)$$

In this case the condition of a sliding mode existence is also a condition of the estimation of the rotor angular position $\Delta\Gamma$ and the angular speed Ω of the synchronous reluctance motor, on the measured stator current and voltage. From equation (6.32), it follows that it can be formulated as an inequality to zero of the stator current modulus. This condition essentially differs from the estimation condition of mechanical variables for permanent magnet *nonsalient-pole* synchronous motor, received in section 6.2. In that case, the mechanical variable estimation, using the stator electrical variables could be possible only by nonzero induction EMF, i.e. by the rotating rotor. The available distinction could be explained by an asymmetry of an air gap of the salience-pole (interior) synchronous motor, due to new possibilities by mechanical variables estimation.

It is obvious that the offered model (6.28) or (6.33) will describe completely the electromagnetic processes in the real synchronous engine (6.26), in case that the error by estimation of the rotor angular position is equal to zero ($\Delta\Gamma = 0$). In this case, the equivalent value of the correcting control μ_1 will be equal to zero, and the equivalent value of the correcting control μ_2 will represent an estimation of the angular speed $\hat{\Omega}$.

To ensure that the used model of the electromagnetic part (6.28), (6.33) is supplemented by the mechanical model with the identification of initial conditions, we should obtain deviations of the estimated angle for subsequent corrections. This model uses estimated values of the rotor angular position error as an input signal. In addition, entered correction should not lead to infringement of existence conditions of sliding motion (6.30), in the assumption that the implementation estimates the angular deviation $\Delta\Gamma$ and the rotor speed $\hat{\Omega}$.

The model of rotor mechanical movement includes the equations of a mechanical part of the motor (1.1) added by the load model. As a first approximation, the load torque could be considered as constant with an unknown value, described by the following equation:

$$\frac{dM}{dt} = 0 \tag{6.36}$$

In this case, the mechanical part of the synchronous reluctance motor is described by the linear vector equation with scalar control M_{el}:

$$\frac{d}{dt}\begin{vmatrix} \Gamma \\ \Omega \\ M \end{vmatrix} = \begin{vmatrix} 0 & 1 & 0 \\ 0 & 0 & -1/J \\ 0 & 0 & 0 \end{vmatrix}\begin{vmatrix} \Gamma \\ \Omega \\ M \end{vmatrix} + \begin{vmatrix} 0 \\ 1/J \\ 0 \end{vmatrix}M_{el} \tag{6.37}$$

To obtain estimates of the state vector components, we can take advantage of the known results on the design of the linear state observer (Luenberger, 1966), (Kwakernaak & Sirvan, 1972), (Leonhard, 2001). On the first stage, a mathematical model of the mechanical part of the synchronous motor (6.37) should be checked for observability, i.e. it is possible to obtain information about the non-measured components of the plant state vector under the available output information. In this case, as output, we consider the rotor angular position Γ. Then the equation of the output variables for the mechanical part of the synchronous reluctance motor can be written as follows

$$\Gamma = |1 \quad 0 \quad 0| \begin{vmatrix} \Gamma \\ \Omega \\ M \end{vmatrix} \tag{6.38}$$

The state vector of the presented above linear system (6.37), (6.38) can be estimated if the observation matrix has a rank equal to the rank of the initial dynamic system, which is three in this case.

The observation matrix for the investigated system looks like:

$$\begin{vmatrix} 1 & 0 & 0 \\ 0 & 1 & 0 \\ 0 & 0 & -1/J \end{vmatrix} \tag{6.39}$$

The determinant of the observation matrix (6.39) is unequal to zero, hence, the observation matrix rank is equal to 3, i.e. the initial dynamic plant (the mechanical part of the synchronous reluctance motor (6.37) with the measured rotor angular position (6.38), is an observable control plant, and there is a possibility to obtain the estimate of the angular speed $\hat{\Omega}$).

According to a design approach to linear full order observers, the observer of the mechanical variables will be described by the following vector of differential equations:

$$\frac{d}{dt} \begin{vmatrix} \hat{\Gamma} \\ \hat{\Omega} \\ \hat{M} \end{vmatrix} = \begin{vmatrix} 0 & 1 & 0 \\ 0 & 0 & -1/J \\ 0 & 0 & 0 \end{vmatrix} \begin{vmatrix} \hat{\Gamma} \\ \hat{\Omega} \\ \hat{M} \end{vmatrix} + \begin{vmatrix} 0 \\ 1/J \\ 0 \end{vmatrix} M_{el} + \begin{vmatrix} l_1 \\ l_2 \\ l_3 \end{vmatrix} \Delta\Gamma \tag{6.40}$$

where l_1, l_2, l_3 are the correctional coefficients selected from a condition of maintenance of desirable zeroing character of the rotor angular position error.

The electromagnetic torque M_{el} of the synchronous reluctance motor, seen as an input, is calculated under the assumption that the estimation errors of the angle $\Delta\Gamma$ on measurements of stator currents and voltages are small.

$$M_{el} = (L_d - L_q)[i_d^* i_q^* + (i_q^*)^2 - (i_d^*)^2]\Delta\Gamma \tag{6.41}$$

As a correcting control in the observer of mechanical variables (6.40) the estimated rotor angular position error $\Delta\Gamma$, obtained as an output of an electromagnetic part model (6.28) or (6.33) is used.

It is obvious that the estimations of mechanical variables obtained by means of the observer will tend to actual values of these variables, since according to (6.37) and (6.40), the vector equation describing the change of an error of estimation, looks like

$$\frac{d}{dt}\begin{vmatrix}\Delta\Gamma\\\Delta\Omega\\\Delta M\end{vmatrix} = \begin{vmatrix}l_1 & 1 & 0\\l_2 & 0 & -1/J\\l_3 & 0 & 0\end{vmatrix}\begin{vmatrix}\Delta\Gamma\\\Delta\Omega\\\Delta M\end{vmatrix} \qquad (6.42)$$

In this case, a characteristic matrix of the vector equation (6.42) is

$$\begin{vmatrix}\lambda - l_1 & 1 & 0\\l_2 & \lambda & -1/J\\l_3 & 0 & \lambda\end{vmatrix} \qquad (6.43)$$

and the characteristic equation is

$$\lambda^3 - l_1\lambda^2 - l_2\lambda - l_3/J = 0 \qquad (6.44)$$

By a corresponding selection of the correctional coefficients l_1, l_2, l_3 there is a possibility to appoint corresponding roots of the characteristic equation (a principle of modal control), located in the left semi plane of a complex plane and the zeroing of mechanical variables errors providing the desirable character.

A possible structure of the observer of mechanical variables for the synchronous reluctance motor is presented in figure 6.2.

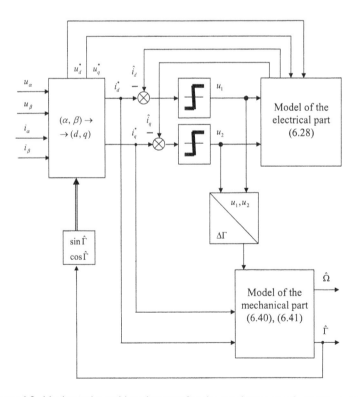

Figure 6.2 Mechanical variables observer for the synchronous reluctance motor

6.3.2 The simplified observer

It is known that rates of the electromagnetic and electromechanical processes proceeding in the synchronous machine essentially differ from each other. Therefore when calculating values of mechanical variables, it is possible to consider the electromagnetic processes established. Under this assumption, the behavior of the synchronous reluctance machine is described by the following system of equations:

$$-ri_d + L_q \Omega i_q + u_d = 0$$
$$-ri_q - L_q \Omega i_d + u_q = 0 \tag{6.45}$$
$$\frac{d\Gamma}{dt} = \Omega$$

The estimated variables, assuming a small error of the rotor angular position estimation $\Delta\Gamma = \Gamma - \hat{\Gamma}$ and taking into account (6.2), are related to the measured electric variable by the following algebraic equations:

$$-ri_d^* + u_d^* + \Omega L_q i_q^* - (-ri_q^* + u_q^* + \Omega L_q i_q^*)\Delta\Gamma = 0 \tag{6.46}$$
$$-ri_q^* + u_q^* + \Omega L_d i_d^* - (-ri_d^* + u_d^* + \Omega L_d i_d^*)\Delta\Gamma = 0 \tag{6.47}$$

Using these algebraic connections between the measured electric variables and the estimated mechanical ones, it is possible to calculate values of the rotor angular position error

$$\Delta\Gamma = \frac{L_d(-ri_d^* + u_d^*)i_d^* + L_q(-ri_q^* + u_q^*)i_q^*}{(L_d - L_q)[(-ri_d^* + u_d^*)i_q^* + (-ri_q^* + u_q^*)i_d^*]} \tag{6.48}$$

and of the angular speed Ω:

$$\Omega = \frac{[L_q(-ri_q^* + u_q^*)i_d^* - L_d(-ri_d^* + u_d^*)i_q^* + L_d(-ri_q^* + u_q^*)i_q^* + L_q(-ri_d^* + u_d^*)i_d^*]}{[L_d(i_q^*)^2 + L_q(i_d^*)^2)]}$$
$$\times \frac{[L_d(-ri_d^* + u_d^*)i_d^* + L_q(-ri_q^* + u_q^*)i_q^*]}{(L_d - L_q)[(-ri_d^* + u_d^*)i_q^* + (-ri_q^* + u_q^*)i_d^*]} \tag{6.49}$$

Given the fact that in the synchronous reluctance machine a longitudinal axis inductance is much larger than the transverse one, i.e. $L_d \gg L_q$ (Leonhard, 2001), (Boldea & Nasar, 2005), (Pahman & Zhou, 2000), an expression for the estimation of the rotor angular position error (6.48) and of the rotor speed, can be somewhat simplified:

$$\Delta\Gamma = \frac{(-ri_d^* + u_d^*)i_d^*}{[(-ri_d^* + u_d^*)i_q^* + (-ri_q^* + u_q^*)i_d^*]} \tag{6.50}$$

$$\Omega = \frac{[-(-ri_d^* + u_d^*)i_q^* + (-ri_q^* + u_q^*)i_q^*](-ri_d^* + u_d^*)i_d^*}{(i_q^*)^2[(-ri_d^* + u_d^*)i_q^* + (-ri_q^* + u_q^*)i_d^*]} \tag{6.51}$$

It is possible to take advantage of the linear observer of mechanical variables (6.40) to filter out the estimations received by means of (6.48)–(6.51).

It was shown that the simplified observer is convenient to use in digital control. Proceeding from the demanded dynamic properties of an electrical drive, the duration of a calculation cycle of control, as a rule, appears large enough in comparison with the time constant of the motor electromagnetic circuit. In this case, neglecting the electromagnetic circuits dynamics appears to be well justified.

REFERENCES

Andreescu, G.D., Popa, A. and Spilca A. "*Sliding mode based observer for sensorless control of PMSM drives – two comparative study cases*". Proc. 7th International Conference on Optimization of Electrical and Electronical Equipment, OPTIM 2000, Brasov, Romania, 2000, CD-ROM.

Boldea, I. and Nasar, S.A. "*Electric drives, 2nd edition*". CRC Press, 2005. 544 p.

Consoli, A. "*Advanced control techniques. Modern Electrical Drives*". Dordrecht, Boston, London: Kluwer Academic Publishers, 2000, pp. 523–582.

Dote, Y. "*Application of modern control techniques to motor control*". Proc. of the IEEE, 1988, vol. 76, no. 4, pp. 438–445.

Holtz, J. "*Sensorless control of induction machines – with or without signal injection*". Proc. 9th International Conference on Optimization of Electrical and Electronical Equipment, OPTIM 2004, Brasov, Romania, 2004, vol. II, pp. XVII–XXXIX.

Kwakernaak, H. and Sivan, R. "*Linear optimal control systems*". New York: John Wiley & Son Inc., 1972. 608 p.

Leonhard, W. "*Control of electrical drives*". Berlin: Springer-Verlag, 2001. 460 p.

Luenberger, D.C. "*Observers for multivariable systems*". IEEE Transactions on Automatic Control, 1966, vol. 11, no. 1, pp. 190–197.

Pahman, M.A. and Zhou, P. "*Interior permanent magnet motors. Modern Electrical Drives*". Dordrecht, Boston, London: Kluwer Academic Publishers, 2000, pp. 115–140.

Ryvkin, S. "*Sliding mode based observer for sensorless synchronous reluctance motor drive*". Proc. International AEGEN Conference on Electrical Machines and Power Electronics, ACEMP'95. Kusadasi, Turkey, 1995, pp. 614–618.

Ryvkin, S.E. "*Sliding mode based observer for sensorless permanent magnet synchronous motor drive*". Proc. 7th International Power Electronics & Motion Control Conference, PEMC'96, Budapest, Hungary, 1996, vol. 2, pp. 558–562.

Ryvkin, S.E. and Izosimov, D.B. "*Algorithm for the identification of mechanical coordinates of an electric drive*". Russian Electrical Engineering (Elektrotechnika), 1994, vol. 65, no. 7, pp. 35–41.

Utkin, V. "*Sliding modes in control and optimization*". Berlin: Springer-Verlag, 1992. 286 p.

Utkin, V., Shi, J. and Gulder, J. "*Sliding Modes in Electromechanical Systems*". London: Taylor & Francis, 1999. 344 p.

Vittek, J. and Dodds, S.J. "*Forced dynamics control of electric drives*". EDIS, Publishing Center of Zilina University, Slovakia, 2003. 356 p.

Chapter 7

Digital control

7.1 MAIN PRINCIPLES OF DIGITAL CONTROL

7.1.1 Features of digital control

From the control viewpoint, transition to digital technologies characterizes itself as a transition to discrete control with level and time sampling. Unlike continuous signals, the input and output signals of digital controls take discrete values during discrete time (Isermann, 1981). Time sampling represents the periodic process characterized by a sampling period T. Without taking into account these features of digital control, direct implementation of analog algorithms leads to the fall of static accuracy of the system, to appearance of oscillatory components with amplitude proportional to the sampling period and to electric dissonance.

Digital control design differs essentially from classical continuous control design. There are many reasons. First, the mathematic models for digital control design are based on ordinary difference equations, which replace the differential equations describing continuous systems. Second, the control system analysis and design are carried out on a sampling period T. This allows control decomposition on processes rates, as equations are simplified at the expense of becoming quasi-constant variables on sampling periods. Third, there is a system memory, in which the values of state vectors and of control ones, during the previous time, are stored, depending on the memory depth m. This information could be used to solve a control problem. Fourth, digital control is often implemented using microprocessors that are limited in speed and processing power. New microprocessors are faster and more powerful, so they tend to overcome these limitations.

The base line by the microprocessor calculation is the computation cycle duration or the calculation step, which is closely related to the measurement time and to the microprocessor processing power. We will assume in further analyses, that the synchronous control principle is used. During the calculation step, which is equal to the sampling period T, the microprocessor solves the control problem, i.e. there is enough processing power to compute the output control during one sampling period. It is necessary to notice that the control problem solving cannot be executed faster, in principle, than two calculation steps. On the first step $[k, k + 1]$, the digital controller, using the available information on variables and references, computes such control for an actuator that by the end of the following step $[k + 1, k + 2]$ the control problem solving will be guaranteed. These controls are actuator inputs on the second step. If the

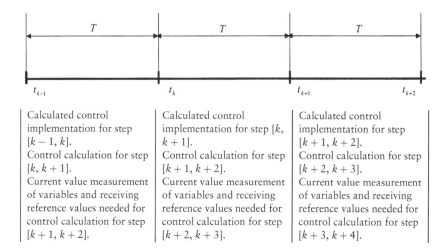

Figure 7.1 Calculations time diagram

specified delay is essential for the given system it should be accepted and compensated in an appropriate way. The resulting time diagram of the calculation organization is shown in figure 7.1.

In the light of the foregoing, for the design of digital control for high quality synchronous drives, it is necessary to develop special methods to design such systems, taking into account and actively using the specified features of digital control.

7.1.2 Digital sliding mode

Transition to digital technologies has raised a question about the potential existence of sliding movement in digital systems. It is necessary to understand what sliding movement is in such systems and its attractive properties. Carrying over analog control to digital control directly leads to deterioration of the static accuracy of the system, the occurrence of oscillatory components (the amplitude being proportional to the period of digitization), and electric dissonance.

Apparently, the first control problem solution regarding synchronous electrical drives was offered by (Baida S. et al., 1988). Then this problem was solved as a digital control system for a robotic electric drive (Isozimov & Skoropad, 1989). It was suggested that the motion arising in the digital system with a following three-zone control was considered under sliding mode.

$$u_i(k) = \begin{cases} U_{\max}(k) & if\ u_{eq}(k) > U_{\max}(k) \\ u_{eq}(k) & \\ U_{\min}(k) & if\ u_{eq}(k) < U_{\min}(k) \end{cases} \tag{7.1}$$

where $U_{max}(k)$, $U_{min}(k)$ are the maximum and minimum values of control that, due to system limitations, can be used on k-th step and $u_{eq}(k)$ is the equivalent control calculated on the k-th step proceeding from a condition of equality to zero of an increment of an control variable error of function $Z_i(k+1) = 0$. The generalization of the given approach for an n-dimensional system is exposed in (Drakunov & Utkin, 1989).

Thus, it is suggested that such motion is to be considered sliding motion in the digital system, when the last will reach the sliding mode manifolds that are the zero value of the control variable error vector $Z(k)$ over two calculation steps (Furuta, 1990), (Utkin et al, 1999). It must be emphasized that the digital sliding mode has only one from the three features of sliding mode, namely the high closed to limiting dynamics.

The existence condition of a digital sliding mode, in this case, looks like:

$$Z(k+1)\bigg|_{u(k)=u_{eq}} = 0 \tag{7.2}$$

7.2 DIGITAL CONTROL DESIGN FOR THE SYNCHRONOUS MOTOR

7.2.1 Synchronous motor difference equations

It is known that the digital control design needs an analytical difference plant model. The difference equations describing the behavior of the control plant are derived from the differential equations of the synchronous motor (1.1), (1.4)–(1.8) under the following assumptions:

- The analysis is carried out on the sampling period T, which is equal to the PWM period;
- The sampling period T is small in comparison with the electric time constant of the synchronous motor;
- The mechanical and magnetic time constants (typically in a range from 10 to 100 milliseconds) and the switching and electrical time constants (typically in a range from 10 to 100 microseconds) of a synchronous motor differ significantly from each other;
- The feed voltage vector of the synchronous motor in coordinates (d, q) on the calculation step T is a constant. Its amplitude is constant as the average value of a converter output vector is a constant over a period PWM T. The average value on period PWM of a phase angle in the coordinate system (d, q) is used. If a more exact model is needed, it could be obtained based on the corresponding integral of convolution;
- The angular speed Ω is a constant (quasi-static changing) parameter in the right part of the equations (1.4)–(1.8) and is equal to the average value on period T;
- The value of the electromagnetic torque M_{eq} in the mechanical variables equation (1.1) is constant, and it is equal to the average value M_{eq} calculated on the elementary ratio of linear change of the torque on the considered interval. (If a more exact model is needed, it could be obtained based on the corresponding integral of convolution);

- The load torque M on period T is constant;
- The synchronous motor is bipolar, i.e. the electric and mechanical angular coordinates are equal.

Under these assumptions, the mechanical processes are very slow in comparison with the sampling period T. In other words, mechanical variables change only slightly during calculations. That allows separately considering the current equations, in which the angular speed Ω will be considered constant (quasi-static changing parameter), and the mechanical variable equations.

As it was emphasized in section 1.1.1, the equation of mechanical movement (1.1) is a general equation for all electrical drives. However the equations of electric balance based on the Kirchhoff second law, and the equation of the electromagnetic torque are defined, as we know, by a type of electrical motor, electric and magnetic circuits and the physical processes proceeding in them. Therefore, they are considered separately for each type of synchronous motor. This applies to a great extent to the difference equations of electrical motors.

The difference equations describing mechanical movement of the synchronous motor, taking into account the assumptions above, could be obtained from the differential equation (1.1) using the transitive matrix

$$\Omega(k+1) = \Omega(k) + \frac{T}{J}[M_{eq}(k) - M(k)] \tag{7.3}$$

$$\Gamma(k+1) = \Gamma(k) + T\Omega(k) + \frac{T^2}{2J}[M_{eq}(k) - M(k)] \tag{7.4}$$

$$M(k+1) = M(k) \tag{7.5}$$

According to the simplifications made the value of the angular speed on k-th interval is defined by the equality:

$$\Omega_{eq}(k) = \Omega(k) + (T/2)(1/J)[M_{eq}(k) - M(k)] \tag{7.6}$$

The values of the voltage vector components in the (d, q) rotating coordinate system are constant and equal

$$\begin{bmatrix} u_d(k) \\ u_q(k) \end{bmatrix} = \begin{bmatrix} \cos\Gamma_{eq}(k) & \sin\Gamma_{eq}(k) \\ -\sin\Gamma_{eq}(k) & \cos\Gamma_{eq}(k) \end{bmatrix} \begin{bmatrix} A(k)\cos\phi(k) \\ A(k)\cos\phi(k) \end{bmatrix} \tag{7.7}$$

where $\Gamma_{eq}(k)$ is the average value of the rotor angular position calculated as

$$\Gamma_{eq}(k) = \Gamma(k) + (T/2)\Omega(k) + (T^2/4)(1/J)[M_{eq}(k) - M(k)] \tag{7.8}$$

where $A(k)$, $\phi(k)$ are the reference value of the amplitude and phase angle of a synchronous motor fed voltage vector, i.e. the inputs of the PWM system.

Considering that the period T is small compared with the electromagnetic time constants of the synchronous motors, transition to the difference equations of electric balance can be essentially simplified by use of Euler approach (Korn & Korn, 2000). In this case, the various types of synchronous motors are described according to their difference equations.

Salient-pole synchronous motor with an excitation winding

$$i_d(k+1) = i_d(k) + \frac{T}{L_1^2}[-L_f r i_d(k) + L_{df} r_f i_f(k) + L_f L_q \Omega_{eq}(k) i_q(k)]$$

$$+ \frac{TL_f}{L_1^2} u_d(k) - \frac{TL_{df}}{L_1^2} u_f(k),$$

$$i_q(k+1) = i_q(k) + \frac{T}{L_q}[-r i_q(k) - L_d \Omega_{eq} i_d(k) - L_{df} \Omega_{eq} i_f(k)] + \frac{T}{L_q} u_q(k),$$

$$i_f(k+1) = i_f(k) + \frac{T}{L_1^2}[-L_d r_f i_f(k) + L_{df} r i_d(k) - L_{df} L_q \Omega_{eq} i_q(k)]$$ (7.9)

$$+ \frac{TL_d}{L_1^2} u_f(k) - \frac{TL_{df}}{L_1^2} u_d(k),$$

$$M_{el}(k) = [L_{df} i_f(k) + (L_d - L_q) i_d(k)] i_q(k)$$

Permanent magnet salient-pole synchronous motor

$$i_d(k+1) = i_d(k) + \frac{T}{L_d}[-r i_d(k) + L_q \Omega_{eq}(k) i_q(k)] + \frac{T}{L_d} u_d(k),$$

$$i_q(k+1) = i_q(k) + \frac{T}{L_q}[(-r i_q(k) - L_d \Omega_{eq}(k) i_d(k) - \Psi_f \Omega_{eq}(k)] + \frac{T}{L_q} u_q(k),$$ (7.10)

$$M_{el}(k) = [\Psi_f + (L_d - L_q) i_d(k)] i_q(k)$$

Synchronous reluctance motor

$$i_d(k+1) = i_d(k) + \frac{T}{L_d}[-r i_d(k) + L_q \Omega_{eq}(k) i_q(k)] + \frac{T}{L_d} u_d(k),$$

$$i_q(k+1) = i_q(k) + \frac{T}{L_q}[-r i_q(k) - L_d \Omega_{eq}(k) i_d(k)] + \frac{T}{L_q} u_q(k),$$ (7.11)

$$M_{el}(k) = (L_d - L_q) i_d(k) i_q(k)$$

Nonsalient-pole synchronous motor with an excitation winding

$$i_d(k+1) = i_d(k) + \frac{T}{L_1^2}[-L_f r i_d(k) + L_{df} r_f i_f(k) + L_f L \Omega_{eq}(k) i_q(k)]$$

$$+ \frac{TL_f}{L_1^2} u_d(k) - \frac{TL_{df}}{L_1^2} u_f(k),$$

$$i_q(k+1) = i_q(k) + \frac{T}{L}[-r i_q(k) - L \Omega_{eq}(k) i_d(k) - L_{df} \Omega_{eq}(k) i_f(k)] + \frac{T}{L} u_q(k),$$ (7.12)

$$i_f(k) = i_f(k) + \frac{T}{L_1^2}[-L r_f i_f(k) + L_{df} r i_d(k) - L_{df} L_q \Omega_{eq}(k) i_q(k)]$$

$$+ \frac{TL_f}{L_1^2} u_f(k) - \frac{TL_{df}}{L_1^2} u_d(k),$$

$$M_{el}(k) = L_{df} i_f(k) i_q(k)$$

Permanent magnet nonsalient-pole synchronous motor

$$i_d(k+1) = i_d(k) + \frac{T}{L}[-ri_d(k) + L\Omega_{eq}(k)i_q(k)] + \frac{T}{L}u_d(k),$$

$$i_q(k+1) = i_q(k) + \frac{T}{L}[-ri_q(k) - L\Omega_{eq}(k)i_d(k) - \Psi_f\Omega_{eq}(k)] + \frac{1}{L}u_q(k), \quad (7.13)$$

$$M_{el}(k) = \Psi_f i_q(k)$$

The average value of the electromagnetic torque $M_{eq}(k)$ is estimated as an arithmetic average of the torque values at the beginning and at the end of a calculation cycle received by using the information about the current components measured at the same time.

7.2.2 Angular speed control

Let us define the operated variable $z(t)$ like this $z = \Omega + cE$, where E is an acceleration. In this case a control variable error $Z(t) = z_z - z$ characterizes a deviation of current value of the control variable from its reference one $z_z = \Omega_z + cE_z$. The zeroing of the control error will provide an exponential reduction of the angular speed error with a time constant $\tau = c^{-1}$. When $c = 0$, there is a final step control, i.e. a digital sliding mode takes place. The asymptotic movement is described by the following equation:

$$\Omega_z(k+2) - \Omega(k+2) = c^2[\Omega_z(k) - \Omega(k)], \quad 0 \le c < 1 \quad (7.14)$$

where $\Delta\Omega(k) = \Omega_z(k) - \Omega(k)$ is the angular speed error.

The digital control of the synchronous motor, providing zeroing of the control variable error $Z(t)$ is constructed by a block principle (Ryvkin et al, 2004), (Ryvkin et al, 2005). It includes:

– The calculation block of the electromagnetic torque reference (a mechanical movement control);
– The current component control.

The reference value of the electromagnetic torque is formed according to the angular speed and position reference values that are electrical drive inputs. It is calculated using the following condition:

$$Z(k+1) = (\Omega_z - \Omega) + c(\Gamma_z - \Gamma) = 0 \quad (7.15)$$

and owing to the mechanical part equations (7.3), (7.5).

The calculated value of the electromagnetic torque is a base to form the reference for a current component control. The transformation proceeding from the electromagnetic torque value to the current components reference ones depends on the synchronous motor type.

As an example, the control for the permanent magnet *nonsalient-pole* synchronous motor will be designed. According to (7.13) the electromagnetic torque value is defined by the product of a permanent magnet flux Ψ_f and the stator current component i_q.

In this case, from the viewpoint of decreasing current loadings of the synchronous motor windings, it is reasonable to make the current component i_d equal to zero, and the current component i_q, equal to the value calculated from the reference value of the electromagnetic torque $M_{eq}(k)$. Using equations (7.11), it is possible to calculate the voltage components values, that are now our control, necessary to solve the control problem

$$u_d(k) = L\Omega_{eq}(k)i_q(k) \qquad (7.16)$$

$$u_q(k) = \frac{M_{eq}(k+1)}{\Psi_f} - i_q(k) - \frac{T}{L}[-ri_q(k) - \Psi_f\Omega_{eq}(k)] \qquad (7.17)$$

If voltage components are calculated and known, and component u_d is calculated so that condition $i_d(k) = 0$ is provided, it is possible to calculate their values for the $(k+2)$-th step:

$$i_q(k+2) = i_q(k) + \frac{T}{L}[-ri_q(k) - \Psi_f\Omega_{eq}(k) + u_q(k)]$$

$$+ \frac{T}{L}\left\{-r\left[i_q(k) + \frac{T}{L}[-ri_q(k) - \Psi_f\Omega_{eq}(k) + u_q(k)]\right]\right.$$

$$\left. - \Psi_f\Omega_{eq}(k+1) + u(k+1)\right\} \qquad (7.18)$$

$$0 = L\Omega_{eq}(k)\left\{i_q(k) + \frac{T}{L}[-ri_q(k) - \Psi_f\Omega_{eq}(k) + u_q(k)]\right\} + u_{dk+1} \qquad (7.19)$$

In this case the electromagnetic torque is equal at the beginning of the $(k+1)$-th step

$$M_{el}(k+1) = \Psi_f\left\{i_q(k) + \frac{T}{L}[-ri_q(k) - \Psi_f\Omega_{eq}(k) + u_q(k)]\right\} \qquad (7.20)$$

and in the end of it, i.e. at the beginning of the $(k+2)$-th step

$$M_{el}(k+2) = \Psi_f\left\{i_q(k) + \frac{T}{L}[-ri_q(k) - \Psi_f\Omega_{eq}(k) + u_q(k)]\right.$$

$$+ \frac{T}{L}\left\{-r\left[i_q(k) + \frac{T}{L}[-ri_q(k) - \Psi_f\Omega_{eq}(k) + u_q(k)]\right]\right. \qquad (7.21)$$

$$\left.\left. - \Psi_f\Omega_{eq}(k+1) + u(k+1)\right\}\right\}$$

The averaged for the $(k + 1)$-th step the electromagnetic torque $M_{eq}(k + 1)$ is calculated them as an average of its arithmetic values at the beginning and the end of this step:

$$M_{eq}(k + 1) = \frac{M_{el}(k + 1) + M_{el}(k + 2)}{2} \qquad (7.22)$$

7.3 DIGITAL DRIVE MECHANICAL VARIABLE ESTIMATION

7.3.1 Problem statement

As it was specified in chapters 3 and 6, it is necessary to have qualitative information both on plant characteristics, and on external influences, for designing a qualitative dynamic plant control. Digital systems, due to its features, such as the presence of memory and the possibility of fast processing of a great volume of information, open ample opportunities on perfecting control for the synchronous electrical drive as a whole. First of all it is related to the improvement of the technical and operational indicators of the synchronous electrical drive at the expense of reducing the number of measured variables and transitions to use by control of estimated (not measured) variables, particularly mechanical ones, and the external perturbations.

As in chapter 6 the known initial information that could be used to receive the variable estimation is a motor current vector I and a motor voltage vector U, or the same actual motor phase currents and voltage, whose measurement does not cause difficulties, and electrical circuit parameters. The observation goal is receiving an estimation of following mechanical coordinates: angular position Γ and rotor speed Ω, accordingly proportional to angle γ_R between motionless axis R of three-phase coordinate system and a mobile axis d and rotor electric frequency ω ($\gamma_R = p\Gamma$; $\omega = p\Omega$, where p is a number of pairs poles of the motor). To simplify matters from now on, it has been assumed that the synchronous motor has two poles, i.e. the electric and mechanical angular coordinates are the same.

The observer providing estimations of unmeasured variables uses a difference imitating model of the investigated plant, whose estimations of the given plant are defined, first of all, by the available information. It is natural that its structure, and also the estimation algorithm of the mechanical coordinates, especially in the case of a nonlinear plant, such as the synchronous motor, essentially depends on the type of the synchronous motor.

7.3.2 Permanent magnet *nonsalient-pole* synchronous motor state observer

The difference equations of the permanent magnet *nonsalient-pole* synchronous motor (7.13) allow us the chance to calculate values of angular speed Ω and angular position Γ by a current memory depth equal to 2. However, the overall system of equations is too complex, both for analysis and design of the estimating system. Therefore, an incremental design procedure for state observers is proposed.

For the permanent magnet *nonsalient-pole* synchronous motor the complex equation on a current vector, taking into account measurements of the current and voltage components are made at the period beginning in motionless coordinates and after they are transformed to the coordinate system connected with a rotor, giving the following equations:

$$e^{j\Gamma_{eq}(k+1)}I(k+1) = \left(1 - \frac{rT}{L}\right)e^{j\Gamma_{eq}(k)}I(k) - jT\Omega_{eq}(k)e^{j\Gamma_{eq}(k)}I(k)$$

$$- j\frac{T}{L}\Psi_f\Omega_{eq}(k) + \frac{T}{L}e^{j\Gamma_{eq}(k)}U(k) \tag{7.23}$$

where $I(k+1)$, $I(k)$, $U(k)$ are complex vectors values of current and voltage in a motionless coordinate system for the two neighboring sampling periods, $e^{j\Gamma_{eq}(k)}$ is a turn matrix and j is the imaginary unit.

Taking into account a small value of the sampling period T, the equation $(e^{j\Gamma_{eq}(k+1)} = e^{j[\Gamma_{eq}(k)+T\Omega_{eq}(k)]} = e^{j\Gamma_{eq}(k)}[1 + jT\Omega_{eq}(k)])$ can be split into two different independent equations (imaginary and real parts), from which it is possible to define two unknown variables $\Gamma_{eq}(k), \Omega_{eq}(k)$:

$$i_\alpha(k+1) - T\Omega_{eq}(k)i_\beta(k+1) = \left(1 - \frac{rT}{L}\right)i_\alpha(k) + T\Omega_{eq}(k)i_\beta(k)$$

$$- \frac{T}{L}\Psi_f\Omega_{eq}(k)\sin\Gamma_{eq}(k) + \frac{T}{L}u_\alpha(k) \tag{7.24}$$

$$i_\beta(k+1) + T\Omega_{eq}(k)i_\alpha(k+1) = \left(1 - \frac{rT}{L}\right)i_\beta(k) - T\Omega_{eq}(k)i_\alpha(k)$$

$$- \frac{T}{L}\Psi_f\Omega_{eq}(k)\cos\Gamma_{eq}(k) + \frac{T}{L}u_\beta(k) \tag{7.25}$$

The angular speed $\Omega_{eq}(k)$ satisfies a quadratic equation, which does not depend on the rotor angular position. (It is necessary to choose one solution from two possible solutions, depending on the rotor rotation direction):

$$\left[\frac{T}{L}\Psi_f\Omega_{eq}(k)\right]^2$$

$$= \left[i_\alpha(k+1) - T\Omega_{eq}(k)i_\beta(k+1) - \left(1 - \frac{rT}{L}\right)i_\alpha(k) - T\Omega_{eq}(k)i_\beta(k) - \frac{T}{L}u_\alpha(k)\right]^2$$

$$+ \left[i_\beta(k+1) + T\Omega_{eq}(k)i_\alpha(k+1) - \left(1 - \frac{rT}{L}\right)i_\beta(k) + T\Omega_{eq}(k)i_\alpha(k) - \frac{T}{L}u_\beta(k)\right]^2 \tag{7.26}$$

Let us notice that equation (7.26) only considers, apart from the defined rotor angular speed, measurements of currents and voltages in a motionless coordinate system and the known synchronous motor parameters.

Using the obtained quadratic equation (7.26) for estimation of angular speed represents certain problems connected not only with the complexity of calculations, but also with the fact that the equations include the measured values of currents and voltage, which, because of presence of noise in measurements, can differ from their true values. Errors of measurements can lead to occurrence of imaginary (complex) solutions of a quadratic equation concerning angular speed (if the quadratic discriminant is negative), which are indisputably unacceptable.

To simplify calculations and excluding imaginary solutions we have to rewrite (7.26) in powers of the sampling period T:

$$
\left[\frac{T}{L} \Psi_f \Omega_{eq}(k) \right]^2
$$

$$
= \left\{ [i_\alpha(k+1) - i_\alpha(k)] + T \left[-\Omega_{eq}(k)[i_\beta(k+1) - i_\beta(k)] + \frac{1}{L}[ri_\alpha(k) - u_\alpha(k)] \right] \right\}^2
$$

$$
+ \left\{ [i_\beta(k+1) - i_\beta(k)] + T \left[\Omega_{eq}(k)[i_\alpha(k+1) + i_\alpha(k)] + \frac{1}{L}[ri_\alpha(k)u_\beta(k)] \right] \right\}^2
$$

$$
\tag{7.27}
$$

The squaring in the right part of equation (7.27) will produce members which do not contain either T, or proportional T or T^2. Due to the small value of the sampling period T, members of the equation (7.27), proportional to T^2, can be neglected, so the equation (7.27) will become the following after some manipulation:

$$
[i_\alpha(k+1) - i_\alpha(k)]^2 + [i_\beta(k+1) - i_\beta(k)]^2
$$

$$
+ \left(\frac{2}{L} \right) T\{[i_\alpha(k+1 - i_\alpha(k)][ri_\alpha(k) - u_\alpha(k)] + [i_\beta(k+1) - i_\beta(k)][ri_\beta(k) - u_\beta(k)]\}
$$

$$
+ 4T\Omega_{eq}(k)[i_\beta(k+1)i_\alpha(k) - i_\beta(k)i_{\alpha 1}(k+1)] = 0
$$

$$
\tag{7.28}
$$

whence angular speed is unequivocally calculated.

$$
\Omega_{eq}(k) = \frac{1}{4T[i_\beta(k)i_{\alpha 1}(k+1) - i_\beta(k+1)i_\alpha(k)]}
$$

$$
\times \left\{ [i_\alpha(k+1) - i_\alpha(k)]^2 + [i_\beta(k+1) - i_\beta(k)]^2 \right.
$$

$$
+ \left(\frac{2}{L} \right) T\{[i_\alpha(k+1 - i_\alpha(k))][ri_\alpha(k) - u_\alpha(k)]
$$

$$
\left. + [i_\beta(k+1) - i_\beta(k)][ri_\beta(k) - u_\beta(k)]\} \right\}
$$

$$
\tag{7.29}
$$

The speed estimation (7.29) does not depend on the flux value Ψ_f. It is probably connected that speed calculation is carried out on a rotation angle of a current vector during a sampling period T. The received estimation of rotor angular speed $\Omega_{eq}(k)$ is based on the digital differentiation of the motor current component change. Therefore, noise will be inevitable, and the use of filters (observers) is required. It is necessary to consider that for estimation quality, the essential influence renders a choice of the sampling period. The smaller the sampling period T, the more influence measurement noise will have. However, on the other hand, value T cannot be too big because of the approached character of the received dependences used condition of relative smallness of the value T. The calculation error owing to the approached character of a ratio (7.29) depends also on values of speed and the electromagnetic torque. The drive simulation defines the compromise value of sampling period T.

Using the speed values from (7.29) equations (7.24), (7.25) for current components could be rewritten as sine and cosine, representing rotor angular position

$$\sin \Gamma_{eq}(k)$$
$$= \frac{-i_\alpha(k+1) + T\Omega_{eq}(k)i_\beta(k+1) + \left(1 - \frac{rT}{L}\right)i_\alpha(k) + T\Omega_{eq}(k)i_\beta(k) + \frac{T}{L}u_\alpha(k)}{\frac{T}{L}\Psi_f\Omega_{eq}(k)}$$

$$(7.30)$$

$$\cos \Gamma_{eq}(k)$$
$$= \frac{-i_\beta(k+1) - T\Omega_{eq}(k)i_\alpha(k+1) + \left(1 - \frac{rT}{L}\right)i_\beta(k) - T\Omega_{eq}(k)i_\alpha(k) + \frac{T}{L}u_\beta(k)}{\frac{T}{L}\Psi_f\Omega_{eq}(k)}$$

$$(7.31)$$

Because of inevitable errors of measurements and calculations, including the received dependences connected with the approached character, the values (7.30), (7.31) are not normalized in the sense that the sum of their squares can differ from one. Therefore, it is necessary either to normalize the received values, or to take advantage of their relation, i.e. $\tan \Gamma_{eq}(k)$:

$$\Gamma_{eq}(k)$$
$$= \arctan \left\{ \frac{-i_\alpha(k+1) + T\Omega_{eq}(k)i_\beta(k+1) + \left(1 - \frac{rT}{L}\right)i_\alpha(k) + T\Omega_{eq}(k)i_\beta(k) + \frac{T}{L}u_\alpha(k)}{-i_\beta(k+1) - T\Omega_{eq}(k)i_\beta(k+1) + \left(1 - \frac{rT}{L}\right)i_\alpha(k) - T\Omega_{eq}(k)i_\beta(k) + \frac{T}{L}u_\beta(k)} \right\}$$

$$(7.32)$$

Thus, by measuring current and voltage stator windings in the synchronous motor, not only the angular speed but also the rotor angular position can be determined (if the direction of rotor rotation is known a priori, or if the rotor rotation direction does not change during the control process). This is due to the presence of two stator windings (in the generalized machine), which transform the synchronous motor into a rotating transformer – a measuring instrument of angular position (to within an electric turn). This property of the synchronous motor, in a combination to digital algorithms, opens up a unique possibility to create a high-precision synchronous electrical drive without mechanical coordinate sensors.

7.3.3   The filter-observer of mechanical variables

As shown in section 7.3.2, digital algorithms allow obtaining estimations of rotor angular speed $\Omega_{eq}(k)$ and angular position $\Gamma_{eq}(k)$, using measurements of currents and voltage. However the received estimations can be noisy, both because of measurement discrepancies and because of discrepancy of the used electromagnetic model (currents change difference models).

Besides, there are not enough available variable estimations for the design of a high quality digital control. In particular, there are no estimations of external disturbing influences, such as the load torque M.

An effective means to solve this problem is the use of digital observers, built on the basis of the difference model of the mechanical variables (7.3)–(7.5). The resulting estimates of angular speed $\Omega_{eq}(k)$ (7.29) and angular position $\Gamma_{eq}(k)$ (7.32) could be regarded as noisy or inaccurate measurements and to use them in the observer of mechanical variables, designed on the basis of the difference model of mechanical processes. The mechanical variables observer is described by the difference equations of the modeling variables. The caret symbol "^" is used as a superscript to differentiate these variables.

$$\hat{\Gamma}(k+1) = \hat{\Gamma}(k) + T\hat{\Omega}(k) + \frac{T^2}{2J}[M_{eq}(k) - \hat{M}(k)]$$
$$+ l_{11}[\hat{\Gamma}(k) - \Gamma_{eq}(k)] + l_{12}[\hat{\Omega}(k) - \Omega_{eq}(k)] \tag{7.33}$$

$$\hat{\Omega}(k+1) = \hat{\Omega}(k) + \frac{T}{J}[M_{eq}(k) - \hat{M}(k)]$$
$$+ l_{21}[\hat{\Gamma}(k) - \Gamma_{eq}(k)] + l_{22}[\hat{\Omega}(k) - \Omega_{eq}(k)] \tag{7.34}$$

$$\hat{M}(k+1) = \hat{M}(k) + l_{31}[\hat{\Gamma}(k) - \Gamma_{eq}(k)] + l_{32}[\hat{\Omega}(k) - \Omega_{eq}(k)] \tag{7.35}$$

Factors l_{ij} used in the mechanical movement variable observer (7.33)–(7.35) are defined by demanded rates estimations of convergence to actual variable values. The choice of these factors could be based on the simulation results taking into account the sensitivity to discrepancy of the motor parameters. The values of $l_{12} = l_{22} = l_{32} = 0$ will be used as initial factor values, i.e. only the angular position estimated by the plant is observed, so obviously, they have to be used. In this case the matrix of own movement (estimation error) is

$$\begin{bmatrix} 1 + l_{11} - \lambda & T & -\dfrac{T^2}{2J} \\[2mm] l_{21} & 1 - \lambda & -\dfrac{T}{J} \\[2mm] l_{31} & 0 & 1 - \lambda \end{bmatrix} \tag{7.36}$$

and the characteristic equation is

$$-\lambda^3 + \lambda^2(3 + l_{11}) - \lambda\left(3 + 2l_{11} + \frac{l_{31}T^2}{2J} - l_{21}T\right) + \left(1 + l_{11} - l_{21}T - \frac{l_{31}T^2}{2J}\right) = 0 \tag{7.37}$$

The desirable characteristic equation (with reference modes) is

$$(\lambda_1 - \lambda)(\lambda_2 - \lambda)(\lambda_3 - \lambda)$$
$$= -\lambda^3 + \lambda^2(\lambda_1 + \lambda_2 + \lambda_3) - \lambda(\lambda_1\lambda_2 + \lambda_1\lambda_3 + \lambda_2\lambda_3) + \lambda_1\lambda_2\lambda_3 = 0 \tag{7.38}$$

In case of observer construction with a final-step convergence, i.e. $\lambda_1 = \lambda_2 = \lambda_3 = 0$, the observer factors are unequivocally defined as follows:

$$l_{11} = -3, \quad l_{21} = -\frac{5}{2T}, \quad l_{31} = \frac{J}{T^2} \tag{7.39}$$

7.4 PARAMETER IDENTIFICATION OF LINEAR DIGITAL SYSTEM WITH VARIABLE FACTORS AND THE LIMITED MEMORY DEPTH

7.4.1 Statement of a parameter identification problem

For the design of a qualitative control for a dynamic plant it is necessary to have qualitative information, both on plant characteristics, and on external influences. In conditions, when characteristics of dynamic plant are not constants, there is a problem of specification and updating of this data for the purpose of the subsequent updating of control. Moreover, the problem of parameter identification can be considered either as an independent problem (Krishnan & Bharadwaj, 1991), or as a part of the general problem of adaptive and robust control (Polyak & Shcherbakov, 2002).

Transition to the digital techniques and, as a consequence, information occurrence in a digital form, not only about current, but also about the previous plant state, is opened by new possibilities on current identification of dynamic plant parameters (Ryvkin, 2007).

Let us consider the linear multidimensional discrete system described by the difference equations with variable factors of type

$$X(k+1) = A(k)X(k) + B(k)U(k) \tag{7.40}$$

where $X \in R^n$ is a state vector; $U \in R^m$ is a control vector; $A(k) = \left\| \begin{smallmatrix} a_1(k) \\ \cdots \\ a_n(k) \end{smallmatrix} \right\|, B(k) = \left\| \begin{smallmatrix} b_1(k) \\ \cdots \\ b_n(k) \end{smallmatrix} \right\|$ are state matrixes of corresponding dimensions; k is the sampling time, in which the system state is considered. The system is completely operated and completely observed. For convenience, we will assume from now on that the sampling periods are all equal, i.e. the sampling period T is constant.

There is a system memory, in which the set of values of state vectors and control appurtenant to the previous sampling points are stored:

$$E = X \cap \Lambda \tag{7.41}$$

where $X = \{X(j)\}$ is a subset of values of the state vector, $j = k_0, \ldots, k+1$, k_0 is the sampling point since which the information on a state vector is stored in memory; $\Lambda = \{U(i)\}$ is a subset of values of a control vector, $i = k_0^*, \ldots, k$, k_0^* is the sampling point, which is the information of a control vector is stored in memory. From now on, it will be assumed for convenience that these sampling points are equal $k_0 = k_0^*$. In this case the control memory depth $m_U = k - k_0 + 1$ is less than the state vector one $m_X = m_U + 1$.

The problem now is to define conditions of factor identification of matrixes $A(k)$, $B(k)$ of the linear multidimensional discrete system (7.40).

7.4.2 Identification condition of matrix factors

The approach is based on the fact that in discrete systems (7.40), (7.41) there is a possibility to use for identification, not only the current values of state variables and control ones, but also their previous values. It is obvious that increasing the memory depth also increases the information volume of state vector and control, i.e. the number of the independent equations increases, allowing defining a larger number of unknown variables.

Two additional rectangular matrixes of the system for definition of the identification condition are entered:

– The system matrix $M(k)$ consisting of a state matrix with the control matrix attached to it

$$M(k) = \|A(k) \quad B(k)\| \tag{7.42}$$

– The information matrix L made of elements of set E. Each line contains the transposed a state vector and the corresponding controls for one sampling point

$$L = \|\mathbf{X}^*\mathbf{U}^*\| \tag{7.43}$$

where $\mathbf{X}^* = \left\| \begin{matrix} X^T(k_0) \\ \cdots \\ X^T(k) \end{matrix} \right\|$ is a matrix of measurements of a state vector, and $\mathbf{U}^* = \left\| \begin{matrix} U^T(k_0) \\ \cdots \\ U^T(k) \end{matrix} \right\|$
is a matrix of measurements of a control vector.

Let us formulate the following statement (an identifiably condition). Factors of state and control matrixes can be defined, if the following conditions are fulfilled:

– The control memory depth is not less than the maximum quantity of nonzero elements in a column of expanded matrix $M(k)$, where $n + m \geq m_U \geq p$ (p is the maximum number of nonzero elements $n + m$).
– Matrix A and B are constant on the memory depth.
– Any square matrix *rank* p allocated from matrix L of dimension $m_U \times (n + m)$ is non-singular.

Proof: Provided that on memory the matrixes A and B are constant, the initial system (7.40) taking into account (7.41) can be rewritten concerning unknown elements of state and control matrixes as follows by using a value set of the n-th component of a state vector, available in memory,

$$X_n^T(k + 1, m_U) = |x_n(1) \quad \cdots \quad x_n(j) \quad \cdots \quad x_n(k + 1)|$$

$$\left| \begin{matrix} X_1(k + 1, m_U) \\ \cdots \\ X_n(k + 1, m_U) \end{matrix} \right| = \left\| \begin{matrix} X^* & 0 & \cdots & 0 & 0 & U^* & 0 & \cdots & 0 & 0 \\ 0 & X^* & \cdots & 0 & 0 & 0 & U^* & \cdots & 0 & 0 \\ \cdots & \cdots & \cdots & \cdots & \cdots & \cdots & \cdots & \cdots & \cdots & \cdots \\ 0 & 0 & \cdots & X^* & 0 & 0 & 0 & \cdots & U^* & 0 \\ 0 & 0 & \cdots & 0 & X^* & 0 & 0 & \cdots & 0 & U^* \end{matrix} \right\| \left| \begin{matrix} a_1^T \\ \cdots \\ a_n^T \\ b_1^T \\ \cdots \\ b_n^T \end{matrix} \right|$$

$$\tag{7.44}$$

It is obvious that the matrix equation (7.44) can be broken on "n" independent equations, consisting of pairs of state and control vector components:

$$|X_l(k + 1, m_X)| = \|X^* \quad U^*\| \quad \begin{vmatrix} a_l^T \\ b_l^T \end{vmatrix} = L \quad \begin{vmatrix} a_l^T \\ b_l^T \end{vmatrix}, \quad l = (1, \dots, n) \tag{7.45}$$

Equation (7.45) will be solvable rather a component of a vector of matrix elements, if in matrix L (7.43) it is possible to allocate a square matrix of rank p. The necessary matrix rank is defined by a maximum number of nonzero elements in a line of incorporated matrix M (7.42).

Proceeding from it, it is possible to define the necessary memory depth to solve the problem. As it was written above a maximum quantity of nonzero elements in a line of an incorporated matrix, i.e. the greatest possible dimension of a vector of matrix elements, is equal to p. Matrix L should have the same rank, in which case the control memory depth cannot be less than p.

It is necessary to consider that the number of matrix equations (7.45) is equal to n, and in each of them an allocated square matrix, depending on defined elements of a matrix, can have the personal set of columns of matrix L (7.43), the maximum number of such square matrixes equals to n. Therefore, in the statement, it is a question of any square matrix made of columns of matrix L (7.43).

Thus, the above resulting statement is proved.

The information about the control memory depth allows formulating a condition on change rates of matrixes elements and the sampling period. If the necessary memory depth to solve the problem is equal to m_U^*, the time interval during which matrix elements should remain constant, is defined as

$$T_c = m_U^* T \tag{7.46}$$

Since the initial plant is continuous, we will take advantage in this case of the theorem of dependence of solving of the differential equations from small parameter. We will assume that we know that the behavior of matrixes elements is described by the differential equations. We will take advantage of the theorem of small factor at a derivative of the higher order (Tikhonov, 1952), (Kokotovic et al, 1976), (Krstic et al, 1995). There are the whole differential matrix equations describing behavior both the control plant and its matrix elements the system. If the last one on the time interval T has a small factor by a higher order derivative the elements of matrixes could be considered as quasi-constants. Moreover, the time constants of the processes proceeding in system, and processes of change of matrix elements, essentially differ from each other. Usually the ratio between time constants is ten. In this case, a condition (7.46) connecting memory, the sampling period, and a time constant of change of matrix elements can be written down as follows:

$$0.1 T_{min} \geq m_U^* T \tag{7.47}$$

where T_{min} is the smallest time constant of change of matrix elements.

7.4.3 Identification of physical parameters

An essential reduction of memory depth can be achieved if we are to identify not the elements of the matrixes of a state and control, but the initial physical parameters, whose changes lead to changes of matrix elements, i.e. the initial matrixes of multidimensional linear discrete system (7.40) are functions of some parameters, and so can be time.

$$A[\alpha(k)] = \begin{Vmatrix} a_1[\alpha(k)] \\ \cdots \\ a_n[\alpha(k)] \end{Vmatrix}, \quad B[\alpha(k)] = \begin{Vmatrix} b_1[\alpha(k)] \\ \cdots \\ b_n[\alpha(k)] \end{Vmatrix} \tag{7.48}$$

where $\alpha(k)$ is a vector of physically changing parameters, $\alpha \in R^r$, $r < n(n+m)$.

In this case, provided that at control memory depth a vector of physically changing parameters is constant and its additive components are included in the entries of the initial system (7.40), the latter, taking into account (7.41), can be rewritten using (7.48) for the vector of physically changing variables using the vector $X_n(k+1, m_U)$.

$$\begin{vmatrix} X_1(k+1, m_U) \\ \cdots \\ X_n(k+1, m_U) \end{vmatrix} = \|N\| \alpha \tag{7.49}$$

where the structure and elements of matrix N depend on a kind of matrixes $A[\alpha(k)]$, $B[\alpha(k)]$ (7.48). In case of multiple occurrences of the vector component, a problem can be the reduction to the above stated kind by transition to small deviations from the rating value. There was a possibility of defining a component of vector α. It is a necessary and sufficient condition for matrix N to be possible to allocate a square matrix of rank k. This statement is easily proved similarly to an identification condition of section 7.2.2. In this case the control memory depth, rates of change of elements of a vector of physical parameters of the plant and the sampling period also are connected by conditions (7.46) and (7.47).

7.4.4 Moment of inertia identification

The solution of an identification problem of the moment of inertia, which is very real for the complex electromechanical systems with a gearless drive, has been presented as an example of use above the specified approach (Ryvkin et al, 2006).

Mechanical movement of electrical drive target shaft is described by the difference equations (7.3)–(7.5).

With a state memory depth $m_X = 2$ and assuming that on this memory depth the load torque and the inertia moment are constant (or considered constant because of slow changes), so.

$$J(k-1) = J(k) \tag{7.50}$$

The moment of inertia, using equations (7.4), (7.5) and (7.50), is calculated as follows:

$$J(k) = \frac{T[M_{el}(k) - M_{el}(k-1)]}{\Omega(k+1) - 2\Omega(k) + \Omega(k-1)} \tag{7.51}$$

Thus, the identifiably condition of the moment of inertia consists in an inequality to zero of a denominator and numerator of expression (7.51):

$$\Omega(k+1) - 2\Omega(k) + \Omega(k-1) \neq 0 \qquad (7.52)$$

$$M_{el}(k) - M_{el}(k-1) \neq 0 \qquad (7.53)$$

Thus the condition of a constancy of the moment of inertia (7.50) and the load torque (7.5) should be met.

It is necessary to notice that in this case there is no necessity to have the information on the load torque value; it is enough to know only its change character.

As control memory depth is known, it can be possible to estimate the sampling period, i.e. the frequency with which the measuring devices are interrogated. Taking into account the time constants of change of the moment of inertia, and the load torque and a condition (7.47) the sampling period is defined as follows:

$$0.05 T_{min} \geq T \qquad (7.54)$$

where $T_{min} = \min\{T_J, T_H\}$, T_J is the time constant of change of the moment of inertia, and T_H is a time constant of change of the load torque.

Thus, at small values of expression (7.52) or (7.53) the definition of the moment of inertia hindered that is obvious from the physical viewpoint. Definition of the moment of inertia is possible either in dynamic operating modes or with use of a test variable signal in the system.

7.5 THE REFERENCE RATE LIMITER

7.5.1 The general problem statement

Designing an effective digital control of a synchronous electrical drive providing high quality indicators becomes considerably complicated, due to the presence of myriad nonlinearities inherent in it. These nonlinearities can be divided into two big classes, depending on the type of their display:

– Nonlinearities caused by the physical nature of the control plant elements;
– Limitation in the range of the system variables.

The AC motor nonlinearity, converter one, backlash (dead space) of an actuation mechanism, dry friction, limitation on the speed of working off an control signal owing to features of realization of the digital control, described in section 7.1.1, concern the first class.

The second class includes the following limitations on system variables:

– On the rotor angular speed, owing to limitations of output voltage of the semiconductor power converter and mechanical durability of the motor and actuation mechanism;

- On the motor electromagnetic torque, owing to limitation of output current of semiconductor power converter and mechanical durability of the motor and an actuating mechanism;
- On the dynamics of an actuating mechanism, owing to technology requirements.

Nonlinearities of the first class are always present in the system and should be considered at control design. Nonlinearities of the second class only exist on certain operating modes of the electrical drive. As their occurrence is related, as a rule, to a control error value in the closed loop for the analysis of system behavior, it is useful to enter the concepts of small and large control errors. A small error is one, which does not become apparent with system variable limitations. On the contrary, a large error leads to variable limitation display. In case we need to design a high dynamic system, it is necessary to make a stability analysis and then a control design with each error separately, including bifurcation points. The specified problem solving above is, especially in the case of a big error, complex enough and labor consuming, leading to the design of a variable structure control. If we are to take only one kind of error into account for control design, e.g. a large error, the system behavior under small errors will be far from optimum on dynamic indicators. Otherwise under a large error, instability of system is possible. It is necessary to notice that in most cases the basic operating mode for an electrical drive is the small error mode.

A possible way for quality operation of electrical drive is the exception mode to limitations (Ryvkin et al, 2004), (Ryvkin et al, 2005). It can be achieved by control formation so that by the large error the system was on the limitation border. In this case, the system uses the maximum of power possibilities of the electrical drive and its dynamic behavior maximizes the energy opportunities. Moreover, such reference rate limiter must meet the following requirements:

- Output of the reference rate limiter that is a control input must be realizable, if the initial input reference is not realized.
- Absence of a dynamic error by the realizable input reference.
- Guarantee that system works in a steady linear zone on deviations of control variables.

7.6 REFERENCE RATE LIMITER

The reference rate limiter design is based on the equation of mechanical movement (1.1) in space of derivatives concerning the angular position Γ, added with bounds on the following mechanical variables of a synchronous motor:

- on acceleration:

$$-\frac{M_{\max} + M}{J} = E_{\min} < E < E_{\max} = \frac{M_{\max} - M}{J} \tag{7.55}$$

where $E_{\max}$ and $E_{\min}$ are the maximum and minimum values of acceleration caused accordingly a maximal and minimal permissible values of electromagnetic torque $|M_{el}| \leq M_{\max}, M_{\max} > 0$;

– on angular speed

$$|\Omega| \leq \Omega_{max} \tag{7.56}$$

where $\Omega_{max} > 0$ is the maximal permissible value of angular speed.

With the account above told, reference rate limiter should represent the dynamic system, considering bounds of the position reference and its rate. The errors between the outputs of the reference rate limiter and the references satisfy the following equations:

$$\frac{d\Delta\Gamma_z}{dt} = \Delta\Omega_z \tag{7.57}$$

$$\frac{d\Delta\Omega_z}{dt} = \Delta E_z \tag{7.58}$$

$$\frac{d\Delta E_z}{dt} = v \tag{7.59}$$

where the low index "z" indicates references values. The low index "lim" have the outputs of the reference rate limiter; $\Delta\Gamma_z = \Gamma_{lim} - \Gamma_z$ is an error between the rotor position output of the reference rate limiter and the rotor position reference; $\Delta\Omega_z = \Omega_{lim} - \Omega_z$ is an error between the rotor angular speed output of the reference rate limiter and the rotor angular speed reference; $\Delta E_z = E_{lim} - E_z$ is an error between the rotor acceleration output of the reference rate limiter and the rotor acceleration reference; v is the reference rate limiter control.

Thus, the design problem of reference rate limiter consists in a selection of a limited value of control v that provides convergence to zero of the errors between the outputs of the reference rate limiter and the references. The additional conditions are the rotor position and the acceleration outputs of the reference rate limiter must be bounded by their maximal values (7.55) and (7.56) accordingly. Additional requirements are not made to inputs of the rate limiter, i.e. to the drive reference. In particular, e.g. the position reference could change in steps, which is characteristic in positioning.

Sliding mode organizing on a following surface could solve the follow-up problem for realized references:

$$S_5 = \Delta E_z + b_1 \Delta\Omega_z + b_2 \Delta\Gamma_z = 0 \tag{7.60}$$

The nature of committing the tracking error $\Delta\Gamma_z$ to zero after occurrence of sliding mode on the surface $S_5 = 0$ depends on the choice of the coefficients b_1, b_2. The selection criterion may be monotony of error change, for instance, which is provided by the real roots of the corresponding characteristic equation.

Since we are interested in the design of a digital controller, operating in discrete time with a sampling period T, we must rewrite the differential equations of the reference rate limiter (7.57)–(7.59) in difference form, assuming that for the sampling period T, the control v value is a constant:

$$\Delta\Gamma_z(k+1) = \Delta\Gamma_z(k) + \Delta\Omega_z(k)T + \frac{\Delta E_z(k)T^2}{2} + \frac{v(k)T^3}{6} \tag{7.61}$$

$$\Delta\Omega_z(k+1) = \Delta\Omega_z(k) + \Delta E_z(k)T + \frac{v(k)T^2}{2} \tag{7.62}$$

$$\Delta E_z(k+1) = \Delta E_z(k) + v(k)T \tag{7.63}$$

A digital sliding mode is organized on surface $S_5 = 0$. It is assumed that on the $(k+1)$-th step provides sliding movement by the selecting the corresponding control $v_5(k)$

$$S_5(k+1) = \Delta E_z(k+1) + b_1\Delta\Omega_z(k+1) + b_2\Delta\Gamma_z(k+1) = 0 \tag{7.64}$$

In this case the demanded value of control $v_5(k)$ is defined from (7.64) taking into account (7.60)–(7.63) as follows:

$$v_5(k) = -\frac{\Delta E_z(k)\left(1 + b_1T + \frac{b_2T^2}{2}\right) + \Delta\Omega_z(k)(b_1 + b_2T) + \Delta\Gamma_z(k)b_2}{T + \frac{b_1T^2}{2} + \frac{b_2T^3}{6}} \tag{7.65}$$

Organizing additional sliding modes along the surfaces characterizing bounds of the angular speed and acceleration could bind outputs of the reference rate limiter:

$$S_1 = E_{\mathrm{lim}} - E_{\max} = 0 \tag{7.66}$$

$$S_2 = E_{\mathrm{lim}} - E_{\min} = 0 \tag{7.67}$$

$$S_3 = c(\Omega_{\mathrm{lim}} - \Omega_{\max}) + E_{\mathrm{lim}} = 0 \tag{7.68}$$

$$S_4 = c(\Omega_{\mathrm{lim}} + \Omega_{\max}) + E_{\mathrm{lim}} = 0 \tag{7.69}$$

After the beginning of sliding movement on surface $S_3 = 0$ or $S_4 = 0$ the movement of the angular speed output of the reference rate limiter to its maximal permissible value is non-periodic with the time constant defined by the selected value of the coefficient c.

The control value of the reference rate limiter $v(k)$ at each step gets out on the obvious logic: the control values providing sliding modes on all five sliding surfaces (7.60), (7.66) to (7.69) are computed. After that, depending on the sign on the control $v_5(k)$, its module is compared with the least on the module from two controls of the same sign providing bounds. If the control $v_5(k)$ value is greater as the bound control, the bound control is used, otherwise is control $v_5(k)$.

It is necessary to notice that the control $v(k)$ itself can be bounded with maximum $v_{\max}(k)$ and minimum $v_{\min}(k)$ values:

$$-v_{\max}^n < v^n < v_{\max}^n \tag{7.70}$$

Thus on the k-th step the design of the demanded value of control of the reference rate limiter $v(k)$ is carried out as follows:

1. The values of the corresponding controls $v_i(k)$ are needed for performance of the sliding mode existence conditions on each of five sliding surfaces on $(k+1)$-th step are computed (7.60), (7.66)–(7.69):

Surface $S_1(k+1) = 0$:

$$v_1(k) = -\frac{E_{\lim}(k) - E_{\max}(k)}{T} \qquad (7.71)$$

Surface $S_2(k+1) = 0$:

$$v_2(k) = -\frac{E_{\lim}(k) - E_{\min}(k)}{T} \qquad (7.72)$$

Surface $S_3 = 0$:

$$v_3(k) = -\left(\frac{c(\Omega_{\lim}(k) + E_{\lim}(k)T - \Omega_{\max})}{T + \frac{b_1 T^2}{2}} + \Delta E_{\lim}(k)\right) \qquad (7.73)$$

Surface $S_4(k) = 0$:

$$v_4(k) = -\left(\frac{c(\Omega_{\lim}(k) + E_{\lim}(k)T + \Omega_{\max})}{T + \frac{b_1 T^2}{2}} + \Delta E_{\lim}(k)\right) \qquad (7.74)$$

Surface $S_5(k) = 0 - (7.65)$.
2. Selecting the control value $v(k)$ from the control set $v_i(k)$ according to the block diagram presented in figure 7.2.
3. Calculation of outputs values of the reference rate limiter on $(k+1)$-th step, that are the references for the drive control:

$$\Gamma_{\lim}(k+1) = \Gamma_{\lim}(k) + \Omega_{\lim}(k)T + \frac{E_{\lim}(k)T^2}{2} + \frac{v(k)T^3}{6} \qquad (7.75)$$

$$\Omega_{\lim}(k+1) = \Omega_{\lim}(k) + E_{\lim}(k)T + \frac{v(k)T^2}{2} \qquad (7.76)$$

$$E_{\lim}(k+1) = E_{\lim}(k) + v(k)T \qquad (7.77)$$

7.7 DIGITAL CONTROL DESIGN FOR THE ELECTRIC DRIVE WITH ELASTIC CONNECTIONS

7.7.1 Control problem statement

Designing an effective servo control for complex electromechanical power converters with elastic mechanical connections is quite an actual problem. It is connected with increasing requirements to control process quality, in particular, to speed increase and a pass-band up to the frequencies comparable to frequency of a mechanical resonance of elastic mechanical system, increase of requirements to dynamic accuracy. Examples of such control plants are different special plants, robots, including space, modern industrial machine work centers, and others complex electromechanical and mechatronic systems.

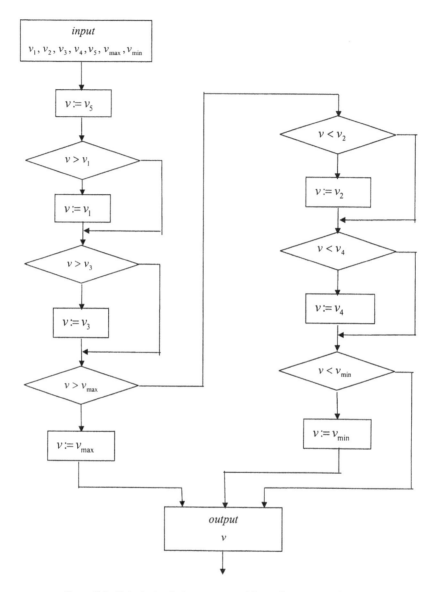

Figure 7.2 Calculating *k*-th step control for reference rate limiter

The presented approach to the digital control design for a mechanical plant with elastic connections is based on the application of a locally distributed digital control (Ryvkin et al, 2003). It is supposed that an electrical drive includes the following elements:

–   A controller;
–   An electric motor (or several electric motors);

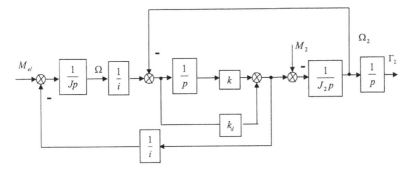

Figure 7.3 Block diagram of the mechanical part of a drive with elastic connections

– A gearbox (with elastic mechanical transmission);
– A final controlled element.

The processes occurring in such control plant are essentially nonlinear due to the change of sign on movement speed of the final control element (backspacing). The control plant is characterized by the presence of some resonant frequencies, which is caused by the distributed masses. The lowest basic frequency of the mechanical resonance lying in the demanded working frequency band of the closed loop position control is usually essential; higher frequencies should be filtered. The control goal is the reference tracing of the angular position the final control element.

Figure 7.3 presents the block diagram of a drive mechanical part. The electric motor with moment of inertia J produces the electromagnetic torque on its shaft M_{el}. The angular speed of the motor shaft is equal to Ω. The reduction factor of mechanical train from a motor shaft the to the final control element (load) is equal to i. Load inertia (the moment of inertia J_2) cooperates with the motor through the elastic mechanical connection with an elasticity factor k and a damping factor k_d. The external disturbing moment M_2 operates on the load. The shaft of inertial load rotates with an angular speed Ω_2. The angular position of the final controlled element is Γ_2.

It is tactical to rewrite the motor parameters in the system connected to the final controlled element for the analysis of the mechanical system. In this case, we have the following mechanical movement equation system of the simplified model:

$$\frac{d\Gamma_1}{dt} = \Omega_1$$

$$\frac{d\Omega_1}{dt} = \frac{1}{J_1}[M_1 - k(\Omega_1 - \Omega_2) - k_d(\Gamma_1 - \Gamma_2)]$$

$$\frac{d\Gamma_2}{dt} = \Omega_2 \tag{7.78}$$

$$\frac{d\Omega_2}{dt} = \frac{1}{J_2}[k(\Omega_1 - \Omega_2) + k_d(\Gamma_1 - \Gamma_2) - M_2]$$

where $\Gamma_1 = \Gamma/i$, $\Omega_1 = \Omega/i$, $M_1 = M_{el}i$, $J_1 = Ji^2$.

Let us analyze the "undisturbed" conditions of elastic fluctuations in the plant. Oscillatory motions in elastic connection will be absent, if relative positions of the first and second masses do not change, i.e. the speeds of both masses coincide:

$$\frac{d(\Gamma_1 - \Gamma_2)}{dt} = \Omega_1 - \Omega_2 = 0$$

$$\frac{d(\Omega_1 - \Omega_2)}{dt} = \frac{1}{J_1}[M_1 - k(\Omega_1 - \Omega_2) - k_d(\Gamma_1 - \Gamma_2)] \qquad (7.79)$$

$$- \frac{1}{J_2}[k(\Omega_1 - \Omega_2) + k_d(\Gamma_1 - \Gamma_2) - M_2] = 0$$

From here a condition of undisturbed movement of the two-mass system is

$$k_d(\Gamma_1 - \Gamma_2) = \frac{J_2 M_1 + J_1 M_2}{J_1 + J_2} \qquad (7.80)$$

This equation (7.80) connects the motor rotor position Γ_1 and the load shaft position Γ_2 in the absence of elastic fluctuations. The class of movements (7.79) that fulfills the condition (7.80) is wide enough. It contains not only movements at zero speed (positioning), but also movement with any constant speed, movement with any constant acceleration (movement on a parabolic trajectory), etc.

Reference values of position and angular speed of the second mass (load) are defined on demanded movement conditions and are equal to $\Gamma_{2z}(t)$ and $\Omega_{2z}(t) = d\Gamma_{2z}(t)/dt$, accordingly. Using the bottom index "z", such reference values of position, angular speed of the first mass and the electromagnetic torque, which satisfy to traffic conditions without elastic fluctuations (7.80) taking into account (7.79), could be written:

$$\Gamma_{1z} = \Gamma_{2z} + \frac{1}{k_d}\left[J_2 \frac{d\Omega_{2z}}{dt} + M_2\right]$$

$$\Omega_{1z} = \Omega_{2z} \qquad (7.81)$$

$$M_{1z} = M_2 + \frac{d\Omega_{2z}}{dt}(J_1 + J_2)$$

7.7.2 Elastic mechanical movement difference model

The difference model of mechanical movement is deduced taking into account that, the variable values needed for the control computing are entered in discrete time. The control values remain constant during sampling period T, which is constant. The initial equation system (7.78) is rewritten using new variables and parameters:

$$N = \Omega_1 - \Omega_2, \quad P = (\Gamma_1 - \Gamma_2), \quad (1/J) = (1/J_1 + 1/J_2),$$

$$N_E = \left(\frac{J_1\Omega_1 + J_2\Omega_2}{J_1 + J_2}\right), \quad P_E = \left(\frac{J_1\Gamma_1 + J_2\Gamma_2}{J_1 + J_2}\right),$$

$$J_E = J_1 + J_2,$$

$$\frac{d}{dt}P = N$$

$$\frac{d}{dt}N = -\left(\frac{1}{J}\right)(kN + k_d P) + \frac{M_1}{J_1} + \frac{M_2}{J_2} \tag{7.82}$$

$$\frac{d}{dt}P_E = N_E$$

$$\frac{d}{dt}N_E = \frac{1}{J_E}(M_1 - M_2) \tag{7.83}$$

The received differential equations system breaks up in two independent subsystems: the first one (7.82) describes the relative movement of two masses, while the second one (7.83) describes the movement of the masses' center.

The system (7.82) describing oscillatory motions, in matrix form, looks like:

$$\frac{d}{dt}\begin{pmatrix} P \\ N \end{pmatrix} = \begin{pmatrix} 0 & 1 \\ -a_1 & -a_2 \end{pmatrix}\begin{pmatrix} P \\ N \end{pmatrix} + \begin{pmatrix} 0 \\ M_i/J_i + M_2/J_2 \end{pmatrix} \tag{7.84}$$

where $a_1 = k_d/J$ and $a_2 = k/J$.

The homogeneous system fundamental matrix is:

$$\Phi(t, 0) = \begin{bmatrix} d_{11} & d_{12} \\ d_{21} & d_{22} \end{bmatrix} \tag{7.85}$$

where $d_{11} = e^{-\sigma T_k}\left(\cos \omega T_k + \frac{\sigma}{\omega}\sin \omega T_k\right)$, $d_{12} = e^{-\sigma T_k}\left(\frac{1}{\omega}\sin \omega T_k\right)$, $d_{21} = e^{-\sigma T_k} \times \left(-\frac{\omega^2 + \sigma^2}{\omega}\sin \omega T_k\right)$, $d_{22} = e^{-\sigma T_k}\left(\cos \omega T_k - \frac{\sigma}{\omega}\sin \omega T_k\right)$, $\omega = \sqrt{a_1 - 0.25 a_2^2}$, $\sigma = 0.5 a_2$ (when ω is real).

Taking into account (7.85) the difference equations of the mass relative movement on the sampling period T are

$$\begin{pmatrix} P(k+1) \\ N(k+1) \end{pmatrix} = \begin{bmatrix} d_{11} & d_{12} \\ d_{21} & d_{22} \end{bmatrix}\begin{pmatrix} P(k) \\ N(k) \end{pmatrix} + \begin{bmatrix} d_{13} \\ d_{23} \end{bmatrix}M(k) \tag{7.86}$$

where $d_{13} = \frac{1}{\sigma^2 + \omega^2}(1 - e^{-\sigma T_k}\cos \omega T_k) - \frac{\sigma}{\omega(\sigma^2 + \omega^2)} \cdot e^{-\sigma T_k}\sin \omega T_k$, $d_{23} = \frac{1}{\omega}e^{-\sigma T_k}\sin \omega T_k$, $M(k) = \frac{M_1}{J_1} + \frac{M_2}{J_2} = constant$.

The difference equations of movement of the masses' center look like:

$$P_E(k+1) = P_E(k) + N_E(k)T + \frac{T^2}{2 J_E}M(k)$$

$$N_E(k+1) = N_E(k) + \frac{T}{J_E}M(k) \tag{7.87}$$

In this case, the difference equations for initial mechanical system (7.78) will be the following:

$$\Gamma_1(k+1)$$

$$= \frac{1}{J_1 + J_2} \left\{ \begin{array}{l} \Gamma_1(k)(J_1 + J_2 d_{11}) + \Gamma_2(k)J_2(1 - d_{11}) + \Omega_1(k)T_k(J_1 + J_2 d_{12}) \\ + \Omega_2(k)T_k J_2(1 - d_{12}) + M(k)\left(\dfrac{T_k^2}{2} + J_2 d_{13}\right) \end{array} \right\}$$

$$\Gamma_2(k+1)$$

$$= \frac{1}{J_1 + J_2} \left\{ \begin{array}{l} \Gamma_1(k)J_1(1 - d_{11}) + \Gamma_2(k)(J_2 + J_1 d_{11}) + \Omega_1(k)T_k J_1(1 - d_{12}) \\ + \Omega_2(k)T_k(J_2 + J_1 d_{12}) + M(k)\left(\dfrac{T_k^2}{2} - J_1 d_{13}\right) \end{array} \right\}$$

$$\Omega_1(k+1)$$

$$= \frac{1}{J_1 + J_2} \left\{ \begin{array}{l} [\Gamma_1(k) - \Gamma_2(k)]J_2 d_{21} + \Omega_1(k)(J_1 + J_2 d_{22}) + \Omega_2(k)J_2(1 - d_{22}) \\ + M(k)(T_k + J_2 d_{23}) \end{array} \right\}$$

$$\Omega_2(k+1)$$

$$= \frac{1}{J_1 + J_2} \left\{ \begin{array}{l} -[\Gamma_1(k) - \Gamma_2(k)]J_1 d_{21} + \Omega_1(k)J_1(1 - d_{22}) + \Omega_2(k)(J_2 + J_1 d_{22}) \\ + M(k)(T_k - J_1 d_{23}) \end{array} \right\}$$

$$(7.88)$$

7.7.3 Digital control design of elastic oscillations

Due to the fact that the rate of regulation process to establish a given value of electromagnetic torque in the electrical drive is considerably higher than the rate of control of mechanical motion, the original problem of control of elastic oscillations can be decomposed into two separate control problems. One problem consists in controlling the electromagnetic torque developed by the electrical drive. The second problem is the actual control of oscillations. There is a control hierarchy, in which for control problem solving of elastic oscillations the electromagnetic torque as a control is used. It allows to substantially simplify the control design and to design separately the electromagnetic torque control and the elastic oscillation one. Each of these controls works with its own sampling period. The sampling period T_k used in digital oscillation control is 10 times the sampling one T used in digital electromagnetic torque control. For the electromagnetic torque control design it is possible to take the above mentioned approach presented in section 7.2.

Load torque changes inevitably lead to change of the relative position of the first and second masses, which are excitation of elastic oscillatory motion. Hence, it is necessary to design a control for its suppression. Control changes once on the sampling period T_k. Considering that value T_k is rather large (in real systems of the order of one millisecond), the control calculation time will be considered occupying a negligible small share of period T_k, (approximately no more than 100 microseconds). In this case, the two-mass system measurements of angular coordinates $\Gamma_1(k)$, $\Gamma_2(k)$, $\Omega_1(k)$, $\Omega_2(k)$ and their reference values $\Gamma_{1z}(k)$, $\Gamma_{2z}(k)$, $\Omega_{1z}(k)$, $\Omega_{2z}(k)$ immediately arrive on an elastic oscillation control. This control instantly gives out the electromagnetic torque reference on the electrical drive. This reference is instantly fulfilled by the electrical drive.

The electromagnetic torque (control) M_1 on k-th step is created by the oscillation control in the form of a sum of a linear combination of variable deviations from their reference values, the value of the load torque and one of the reference acceleration. It is equal to:

$$M_1(k) = \begin{bmatrix} a_1 & a_2 & a_3 & a_4 \end{bmatrix} \cdot \begin{bmatrix} \Gamma_2(k) - \Gamma_{2z}(k) \\ \Gamma_1(k) - \Gamma_{1z}(k) \\ \Omega_2(k) - \Omega_{2z}(k) \\ \Omega_1(k) - \Omega_{1z}(k) \end{bmatrix} + M_2(k) + \frac{d\Omega_{2z}(k)}{dt}(J_1 + J_2) \quad (7.89)$$

where a_1, a_2, a_3, a_4 are the factors of a feedback chosen from a condition of a desirable placing of roots of differential equations (7.88) rewritten concerning control errors.

Thus, for position control of mechanical system with elastic connections it is necessary on k-th step of control computing:

- To know the angular position reference of the second mass $\Gamma_{2z}(k)$;
- To know the speed reference of this one $\Omega_{2z}(k)$;
- To know the reference acceleration of this one $d\Omega_{2z}/dt$;
- To know the value of the load torque M_2;
- On reference values of the position $\Gamma_{2z}(k)$, the angular speed $\Omega_{2z}(k)$ and the accelerations $d\Omega_{2z}/dt$ of the second mass and the load torque M_2 to compute according to (7.81) reference values of the position $\Gamma_{1z}(k)$ and the speeds of the first mass $\Omega_{1z}(k)$ providing movement without elastic oscillation;
- To know the current (measured) values of the angular speeds $\Omega_1(k)$, $\Omega_2(k)$ and the positions $\Gamma_1(k)$, $\Gamma_2(k)$ of the both first and second masses;
- To compute an output of elastic oscillation controller on current values of deviations of positions and speeds of masses from their reference values according to (7.89).

Feedback factors a_1, a_2, a_3, a_4 pay off in advance and are put into the controller. By the factors calculation it is necessary to know parameters of mechanical system.

7.7.4 State variable observer

A number of simplifying assumptions were used at the design of the control plant with the elastic connections. In this case the use of the state variables observer is necessary. The necessity of the observer is due mainly because the plant model (system of mechanical movement) is obviously inexact in the high-frequency area (at frequencies above the basic resonant frequency). The plant is characterized by presence of several resonance frequencies. In these conditions using direct error feedbacks with big coefficients is fraught with serious complications, such as stability loss in the closed control loop. The solution consists in the use of a state variable observer, which allows making decomposition feedback on a frequency sign. In the working area of feedback frequencies, a loop occurs through the control plant, while the high-frequency component of a feedback signal becomes isolated through the state observer, which, unlike the plant, has the reference structural properties (plant sensors outputs are filtered, without a dynamic error).

Thus, the use of observers makes a methodological basis of control design with the big feedback factors. It must be emphasized that discrepancy of the control plant structure (order) in the field of high frequencies does not allow closing the feedback loop, even when using high quality, exact sensors, of adjustable variables.

Let us notice the fact that, in the field of the working frequencies, the observer allows a compensating cumulative discrepancy of the model, connected with a discrepancy of the definition of plant parameters, a discrepancy of working off the electromagnetic torque reference, simplifications by the conclusion of the plant model, etc. It is reached at the expense of use of disturbance model (the observer of not measured disturbance), reflecting discrepancy of knowledge of plant. Besides the observer allows raising simultaneously accuracy of measurements, in particular, to exclude, or it is essential to lower influence of step-type level behavior of measurements, characteristic for digital sensors or pulse transducers, or for sensor analog-digital converters by their final length.

The mechanical movement drive sensors are the sensors of position and angular speed of both the first and the second masses. They assume that the load torque on the sampling period T_k is constant. The digital observer is designed based on the mechanical variables difference model (7.88).

The mechanical variable observer is described by the following difference equations related to modeling variables. Notice that the observed variables have a "^" caret at the top, to distinguish them from the corresponding real variables.

$$\hat{\Gamma}_1(k+1)$$

$$= \frac{1}{J_1+J_2}\left\{\begin{array}{l} \hat{\Gamma}_1(k)(J_1+J_2d_{11}) + \hat{\Gamma}_2(k)J_2(1-d_{11}) + \hat{\Omega}_1(k)T(J_1+J_2d_{12}) \\ +\hat{\Omega}_2(k)TJ_2(1-d_{12}) + M_1(k)\left(\frac{T^2}{2}+J_2d_{13}\right) + \hat{M}_2(k)\left(\frac{T^2}{2}+J_2d_{13}\right) \\ -l_1[\hat{\Gamma}_1(k)-\Gamma_1(k)] \end{array}\right\}$$

$$\hat{\Gamma}_2(k+1)$$

$$= \frac{1}{J_1+J_2}\left\{\begin{array}{l} \hat{\Gamma}_1(k)J_1(1-d_{11}) + \hat{\Gamma}_2(k)(J_2+J_1d_{11}) + \hat{\Omega}_1(k)TJ_1(1-d_{12}) \\ +\hat{\Omega}_2(k)T(J_2+J_1d_{12}) + M_1(k)\left(\frac{T^2}{2}-J_1d_{13}\right) + \hat{M}_2(k)\left(\frac{T^2}{2}-J_1d_{13}\right) \\ -l_2[\hat{\Gamma}_2(k)-\Gamma_2(k)] \end{array}\right\}$$

$$\hat{\Omega}_1(k+1)$$

$$= \frac{1}{J_1+J_2}\left\{\begin{array}{l} [\hat{\Gamma}_1(k)-\hat{\Gamma}_2(k)]J_2d_{21} + \hat{\Omega}_1(k)(J_1+J_2d_{22}) + \hat{\Omega}_2(k)J_2(1-d_{22}) \\ +M_1(k)(T+J_2d_{23}) + \hat{M}_2(k)(T+J_2d_{23}) - l_3[\hat{\Omega}_1(k)-\Omega_1(k)] \end{array}\right\}$$

$$\hat{\Omega}_2(k+1)$$

$$= \frac{1}{J_1+J_2} \left\{ \begin{array}{l} -[\hat{\Gamma}_1(k) - \hat{\Gamma}_2(k)]J_1 d_{21} + \hat{\Omega}_1(k)J_1(1-d_{22}) + \hat{\Omega}_2(k)(J_2+J_1 d_{22}) \\ +M_1(k)(T-J_1 d_{23}) + \hat{M}_2(k)(T-J_1 d_{23}) - l_4[\hat{\Omega}_2(k) - \Omega_2(k)] \end{array} \right\}$$

$$\hat{M}_2(k+1) = \hat{M}_2(k) + l_5[\hat{\Gamma}_2(k) - \Gamma_2(k)] + l_6[\hat{\Omega}_2(k) - \Omega_2(k)] \qquad (7.90)$$

The factors l_i used in the observer of mechanical variables (7.90), as well as in the observer described in section 7.3 are defined by demanded convergence rates of estimation variables to their current values. The choice of these factors resulting from simulation, take into account estimation sensitivity to the discrepancy of motor parameters.

REFERENCES

Baida S., Izosimov D., Ryvkin S. *"Drive digital sliding mode control"* – Abstracts 2nd All-Union Seminars on Robotics and Flexible Manufacturing Systems, Chelyabinsk, 1988, pp.87–88 (in Russian).

Drakunov, S. and Utkin, V. *"On discrete-time sliding modes,"* Proc. Nonlinear Control System Design Conference, Capri, Italy, 1989, pp. 273–278.

Furuta, K. *Sliding mode control of a discrete system.* System and Control Letters, 1990, vol. 14, no. 2, pp. 145–152.

Isermann, R. *Digital control systems.* Berlin: Springer-Verlag, 1981. 566 p.

Izosimov, D.B. and Skoropad, S.V. *"Digital control for robot drive by using sliding mode".* Bulletin of the USSR Academy of Sciences. Technical cybernetics, 1989, no. 1, pp. 146–153 (in Russian).

Kokotovic, P.V., O'Malley, R.B. and Sannuti, P. *"Singular perturbation and reduction in control theory".* Automatica, 1976, no. 12, pp. 123–132.

Korn, G.A. and Korn, T.M. *Mathematical handbook for scientists and engineers.* USA: Dover Publications, 2000. 1152 p.

Krishnan, R. and Bharadwaj, A.S. *"A review of parameter sensitivity and adaptation in indirect vector controlled induction motor drive systems".* IEEE Transactions on Power Electronics, 1991, vol. 6, no. 4, pp. 695–703.

Krstic, M., Kanellakopoulos, I. and Kokotovic, P. *Nonlinear and Adaptive Control Design.* New York: Wiley, 1995. 563 p.

Polyak, B.T. and Shcherbakov, P.S. *Robust stability and control.* Moscow: Nauka, 2002. 303 p. (in Russian).

Ryvkin, S., *"Estimation of state vector components in a nonlinear singular system".* Control Sciences, 2007, no. 4, pp. 8–13 (in Russian).

Ryvkin, S., Izosimov, D., Aksarin, D. and Cernat, M. *"Digital sensorless xontrol of an exterior permanent magnet synchronous motor".* Proc. 11th International Power Electronics & Motion Control Conference, EPE, PEMC 2004, 2004, CD-ROM.

Ryvkin, S., Izosimov, D. and Baida, S. *"Flex mechanics digital control design".* Proc. IEEE International Conference on Industrial Technology (IEEE ICIT'03), 2003, pp. 298–303.

Ryvkin, S., Izosimov, D. and Baida, S. *"Digital reference rate limiter design".* Proc. 9th International Conference on Optimization of Electrical and Electronic Equipment, OPTIM '04, Brasov, Romania, 2004, vol. 3, pp. 103–108.

Ryvkin, S., Izosimov, D. and Palomar-Lever, E. *"Digital sliding mode based references limitation law for sensorless control of an electromechanical system".* Journal of Physics: Conference Series, 2005, no. 23, pp. 192–201.

Ryvkin, S., Izosimov, D. and Vinogradov, A. "*Identification of the moment of inertia in the digital control drive*". Proc. 12th International Power Electronics & Motion Control Conference, EPE, PEMC 2006, Portoroz, Slovenia, 2006, pp. 438–443.

Ryvkin, S., Palomar-Lever, E. and Izosimov, D. "*Digital sliding mode based sensorless control of an electromechanical system*". Proc. 31st Annual Conference of the IEEE Industrial Electronics Society, IECON, 2005, vol. 1, pp. 28–34.

Tikhonov, N. "*Systems of differential equations with a small parameter multiplying derivation*". Mathematicheskii Sbornik, 1952, vol. 73, no. 31, pp. 575–586 (in Russian).

Utkin, V., Shi, J. and Gulder, J. *Sliding Modes in Electromechanical Systems*. London: Taylor & Francis, 1999. 344 p.

Chapter 8

Practical examples of drive control

8.1 HIGH SPEED SYNCHRONOUS DRIVE SENSORLESS CONTROL

8.1.1 Features of the control system

The following work was carried out in cooperation with the Federal State Unitary Enterprise "Research and Production Enterprise Josephian All-Russia Scientific Electromechanic Research Institute" in the frame of the Federal target program "National technological base". The project name is: "Prototyping a small-sized high speed drive for oil drowned pumps for oil extraction with inverter and microprocessor power control up to 200 kW" (Ryvkin et al, 2005).

The goal for this project is to design an electrical drive for oil drown pumps, having better features that those from present drives in the Russian market. There are oil drown pumps from Russian companies such as Alnas, Avanto, Triol, Elekton, and also from some non-Russian manufacturers such as Centrilift (a division of Baker Hughes) and Reda (a division of Schlumberger).

The developed equipment for oil extracting should possess small dimensions, high reliability, and a large period between repairs. Thus it should provide a high quality control in combination with small power consumption at extracting liquids. The basic part of such equipment is the oil drowned pump. It should function reliably with high efficiency in heavy operational conditions. Reliability in this case is understood in every sense. It includes both reliability of pump performance, reliability of the power supply, and also information reliability. There are some ways to increase reliability. One way is increasing reliability by improving the electrical drive. Another one includes the use of new controls, information receiving, observation, processing and transferring, and the new general circuit solutions in the area of various elements connection and their reservation. These two ways to increase reliability supplement each other. The chosen technical solutions essentially affect the electrical drive control requirements.

Assuming the above mentioned, a pump dimension reduction was suggested. A specially designed high-speed synchronous motor was used. It has samarium-cobalt permanent magnets having stable magnetic characteristics at a temperature of 250 degrees Celsius. This allows a reduction in five to six times the length of drowned part of pump electrical drive, in comparison with the existing ones, and it provides a drowned part diameter of about 100 mm. The price to pay for these improvements was a requirements increase for the power source, since the frequency of feeding voltage should be one kHz approximately, and to the control, owing to the small moment of inertia of the motor rotor. In this situation it is disadvantageous to transfer power from the land converter to a powerful synchronous motor in the specified frequencies. A solution was suggested, in which the inverter settles directly near the synchronous motor, i.e. it is the drowned element. The power transmission is carried out from the land rectifier through a direct current link. The voltage value of a direct current link is defined by a drive rated power. In order to reduce current and cable losses at total drive power (200 KW), the voltage of the direct current link is 2,400 volts.

Reliability improvement is carried out by a two level element reservations of the electric drive drowned part. The first level is the electrical drive level. The second one is the level of the electrical drive elements. Four 50 KW complete electrical drives working on one output shaft are used instead of one 200 KW electrical drive at the first reservation level. The complete drive is the finished product, which includes a synchronous motor, VSI and a control. The inputs of VSIs are connected consistently. Such schema allows dividing a high voltage on the input drive to produce the source synchronous motor voltage at common industrial level of 380 volts. It also allows raising synchronous motor reliability and survivability of all systems as a whole. The failing complete drive VSI is shunted and disconnected from the synchronous motor by contactors. In case of malfunction of one of the electrical drives, the voltage in the direct current link automatically decreases to 600 volts, but the other drives remain working normally. An even more substantial contribution to increase inverter reliability is brought by the considerable redundancy on a load current of the IGBT power modules (the second reservation level). Six three-phase bridge modules with rated current 150 amperes in parallel are included in each of the four inverters, each one handling a total current no greater than 20 amperes (at a case temperature of +80 °C). In other words, we can tolerate the failure of three of the six modules. Due to the small loading of the VSI modules, efficiency is considerably increased (near 98%) and heat removal is facilitated. Figure 8.1 shows the simplified scheme of power circuits of the 200 kW electrical drive.

The digital control has a hierarchical distributed structure. The top level is a land control station. The bottom level is the locally distributed control, whose structure

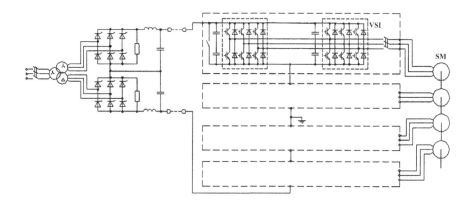

Figure 8.1 Power circuits for 200 kW electrical drive

includes four electrical drive controls. Each of them is constructed based on the TMS320F2810 microprocessor, which controls and makes diagnostics of the electrical drive. Diagnostics signals arrive on the land control station. The top level takes care of the speed reference and VSI connections and disconnections. Thus, the locally distributed drive architecture has a considerable number of transmission channels of power and information. The most critical channels, from a reliability viewpoint, are the information transfer channels. It is necessary to consider issues of electromagnetic compatibility in order to work them out. One possible way to solve this problem is to reduce the amount of information provided by channels and sensors, i.e. to stop using mechanical variable sensors in the drive.

The complete electrical drive control should be high-speed (due to the small rotor moment of inertia). Along with the control problem, we have to solve the problem of estimating the unmeasured mechanical variables of the rotor (angular rotor speed Ω and angular rotor position Γ) and the unmeasured load (load torque M) to fully solve control and observation problems.

The control design is based on breaking up an initial control problem into two independent ones. The first one consists in controlling the output mechanical variables of the electrical drive, and the second one in receiving estimations of the current values of the variables, which are necessary for the control organization. Figure 8.2 presents the control block diagram. The algorithms used in the control system are described in detail in sections 7.2, 7.3, 7.5. Therefore, we will only stop shortly here on the working features of these controls. We note the following facts:

- The frequency of the PWM is fixed;
- The sampling period T is equal or multiple to the PWM period;
- Variable measurements are carried out once at the beginning of the sampling period. In current and voltage measurements at the stator windings the PWM component is absent;

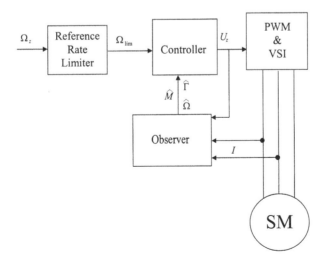

Figure 8.2 Control system structure

- Estimations of mechanical variables are calculated once on each sampling period T;
- Controls are calculated once on the sampling period T.

To eliminate a PWM component, the current measurement must be done in certain moments on period PWM. As a current value of voltage an average voltage value for period PWM could be used, e.g. a reference value.

To estimate angular position, angular speed, and load torque, a two step procedure is used. On the first step of the $(k + 1)$-th calculation period, we use the following information:

- current values of the current vector component $i_\alpha(k + 1)$, $i_\beta(k + 1)$, measured at the period beginning;
- current component values in the previous period $i_\alpha(k)$, $i_\beta(k)$, being in the processor memory;
- components of a stator voltage vector $u_\alpha(k)$, $u_\beta(k)$, calculated and stored earlier on.

Angular position $\Gamma_{eq}(k)$ is estimated using (7.32).

On the second step, this estimation is used as a correcting signal for the observer-filter of mechanical variables constructed according to (7.33)–(7.35). The choice of correction factors of observer output, e.g. estimations of rotor angular position, angular speed and the load torque, is defined by desirable speed of convergence of an estimation error to zero.

In order to avoid problems arising at start-up and a backspacing of a drive owing to big control errors, a reference rate limiter, described in section 7.5, is included in the control. There are usually not enough resources to solve the control task in the above mentioned cases, so the use of the voltage and current limit values is mandatory. There is a big problem here related to changing the system gain. For example, let us suppose that the control tries to produce 100 V, but the power converter can only get 70 V. If the control changes the system structure, the system could become unstable. Another stability analysis must be performed in this case and the new control coefficients must be calculated. As a result, the switching structure system would be complicated in design and use. A different solution is the rate limiter. It always guaranties that the drive will have a reference that can perform this task, so that there would be no problem with these limitations.

The control is designed by the block principle described in section 7.2.

The PWM presented in chapter five is used for VSI switching loss minimization.

8.1.2 Simulation model

The developed control was simulated using the software package MatLab 6.5.0 and Simulink 5.0. The modeling system included the mathematical models of the generalized motor, the VSI, the PWM, the observer, the controller, the rate limiter, and the angular speed master, the load torque master, input blocks of controller and motor parameters and graphic block to display the simulation results. Besides, the model of a direct current link (RLC circuit) was used. That allowed estimating the influence of transients and a "long line" on voltage fluctuations in a direct current link feeding VSI.

Table 8.1 Input data for simulation

Parameters and Units	Synchronous motor	Generalized machine
Phase voltage, U_{ph} (V)	220	270
Phase resistance, R (Ω)	0.0385	0.0385
Phase inductance, L (mH)	0.24	0.24
Poles	2	1
Moment of inertia, J (kg · m^2)	0.01	0.025
Power, P (kW)	50	50
Maximum angular speed, Ω_{max}	30,000 rpm	$1,000 * 2\pi$ rad/sec
Magnetic flux, Ψ_f (Wb)	0.05	0.0606
Rating motor torque, M_r (N · m)	15.9	27.95
Maximum active current, i_{qmax} (A)	215	263

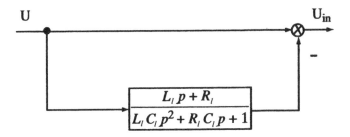

Figure 8.3 Direct current link model

Parameters of the synchronous motor and the corresponding generalized machine are shown in table 8.1.

The models of the inverter (PWM), observers, a controller and the rate limiter work in discrete time. The sampling period (PWM period) was $T = 1/15,000$ sec.

The direct current link model has the following parameters:

1. The line resistance is 0.34 Ω (after recalculation on a one complete drive);
2. The line inductance is 1.5 mH;
3. The capacity C_1 on exit from a feed line is 5,000 µF (on an input of the inverter of one complete drive);
4. The rated voltage value of a direct current link is 600 V (it is 660 V U_r for the generalized machine).

The direct current link model is assembled from Simulink elements. The structure of a direct current link model is shown in figure 8.3.

It is possible to change the references and synchronous motor parameters in the simulation process.

The stability degree and sensitivity of a control system are in many respects defined by the established rates of control in a controller, namely:

− Filtration degrees of the mechanical variables observer;
− Speed of a mechanical movement controller;
− Current controller speed (control rates).

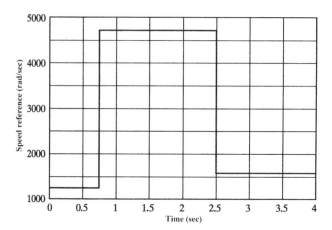

Figure 8.4 Rotor angular speed reference

The observer, the mechanical controller, and the current controller together make the structure similar to the structure of a subordinated control in the closed loop. The "inner" current control loop is captured by an "external" one of mechanical movement, forming a current reference. In turn, the reference for a mechanical movement controller is formed by the observer. (The observer input is formed by inertia-free calculator of angular speed and angular position). It is known that for an optimum course of engineering maintenance of transients (the maximum speed at rather small control), rates of control in internal and external loops should differ approximately in 2–2.5 times. Considering that the observer represents a third order system, we conclude that the rate (time constants) processes in the observer, mechanical movement controllers, and power should be approximately related in these proportions 1:12:30. The following values of the characteristic roots of the differential equations of the observer and the controllers were obtained by simulation, corresponding to the specified ratio of rates

$$\hat{E}_{observer} = 0.995; \quad \hat{E}_{mech\ drive} = 0.96; \quad \hat{E}_d = \hat{E}_q = 0.89$$

The general rates of the control processes were defined by experimentation, using the results of test simulation.

Simulation was done with the use of real physical variables for the generalized machine, measured in SI (Système International d'Unités or International System of Units: volt, ampere, meter, radian, second, etc). The axis of abscissas of the resulting oscillograms corresponds to the time axis in seconds, and the axis of the ordinates represents the physical variable in the SI system. For processes comparability of the analysis of stability, sensitivity to parameters change and others "non ideal facts", inherent in the real system, the model of change of the angular speed reference (figure 8.4) and changes of drive load (figure 8.5) are used. The chosen character of change of angular speed reference allows estimating the motor dynamics dispersal in all the range of possible amplitude changes of step reference, and also motor braking.

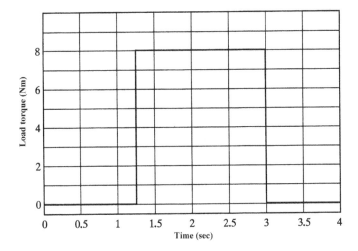

Figure 8.5 Load torque

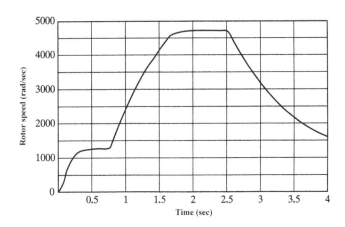

Figure 8.6 Reference rate limiter output and actual rotor angular speed (both curves coincide)

Load surge and dump, during dispersal and braking respectively, allows estimating the drive rigidity in the "heaviest" operating modes.

8.1.3 Drive rating parameters

Figures 8.6 to 8.11 present various processes oscillograms.

– Figure 8.6 shows output of the reference rate limiter, i.e. the rotor angular speed control reference, and the actual angular speed.
– Figure 8.7 shows the actual angular speed and its estimate by the observer.
– Figure 8.8 shows the actual load torque and its estimate by the observer.
– Figure 8.9 shows change of direct current line voltage.

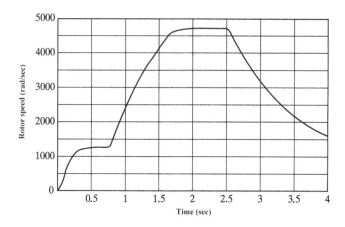

Figure 8.7 Actual and estimated angular speed (both curves practically coincide)

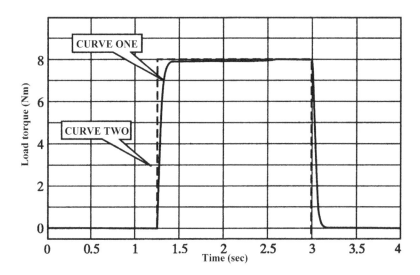

Figure 8.8 Load torque (curve two) and its estimate (curve one)

- Figure 8.10 shows the active current reference and its actual value (characterizes current controller work).
- Figure 8.11 shows change of the actual value of a magnetization (reactive) current (the reference value is zero. It characterizes a current controller work too).

It can be concluded from the simulation results (coincidence of graphs) that if the parameters of the plant and controllers are known, our closed loop system is stable, even if there are no sensors for the angular rotor speed, in the whole range of angular speeds (1:10) and the load torque, and even with bounded feed voltage and current. Stability with bounded variables is essential and especially important in the case of

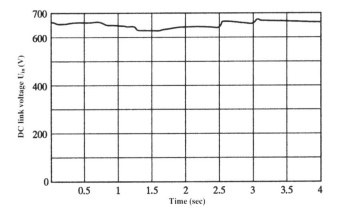

Figure 8.9 Direct current link voltage

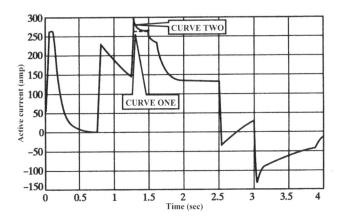

Figure 8.10 Reference (curve one) and actual (curve two) values of the active current

the rated values with big amplitudes. The transients are mainly caused by bounded variables.

There are no harsh requirements for the electrical drive performance of the oil drowned pump. Therefore an acceleration time from zero to the maximum angular speed less than 2 sec is comprehensible (it must be emphasized this is the acceleration time at the nominal load torque). It is possible at slow enough electrical drive dynamics – as it follows from the oscillograms – to receive a voltage change range of a direct current link long line in the order of 620 to 680 volts. Let us notice, that in practice, instant, spasmodic loading changes do not take place, so it is realistic to expect a smaller real range of direct current link voltage. The resonant fluctuations do not increase practically with the chosen nominal parameters of RLC circuit (a long line of a direct current link).

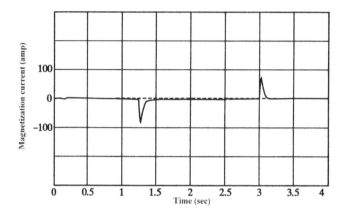

Figure 8.11 Magnetization current (reference value is zero)

The controlled rotor angular speed range in the simulation was bounded with a specified ratio of 1:10, and the start-up of the synchronous motor and short circuit of a control loop performed with zero initial values of rotor angular speed. There are reasons to believe that the actual range is considerably higher, which follows from tracking the small amplitude reference.

Furthermore, let us emphasize that the developed control allows to carry out an estimation of the load torque, both in static and dynamic operating modes (acceleration, braking). It can be quite useful, not only from the viewpoint of control design, but also for electrical drive diagnostics and also the oil drowned pump as a whole.

8.1.4 Sensitivity research to parameter variation

Thanks to the high efficiency of the synchronous motor, the resistance of stator windings does not contribute practically to the formation of the demanded motor feed voltage by the controller. Therefore a change in resistance value does not render practically any influence in the electrical drive transient (certainly, thus the losses rate in the synchronous motor changes considerably). A simulation of an electrical drive at 20% resistance value change has been performed, but it made no difference in the control processes.

A change of inductance winding value of a synchronous motor does not affect the observation practically, and the rotor angular speed control, and also on processes in a direct current link. Some differences are observed in estimation of the load torque and control of active and reactive currents. Figures 8.12 to 8.14 show oscillograms of these processes at 20% increase of the stator winding inductance ($L = 288$ mH).

Figures 8.15 to 8.18 show processes oscillograms at 20% reduction of a magnetic flux ($\Psi_f = 0.048$ Weber). A change in the magnetic flux does not affect practically the estimation of the angular speed; however the influence of this change on angular speed control is noteworthy. It is connected, probably, because the "wrong" reference of a magnetic flux leads to an error in definition of the electromagnetic torque, and as consequence, to an error in estimation of the load torque. It appears from the resulting

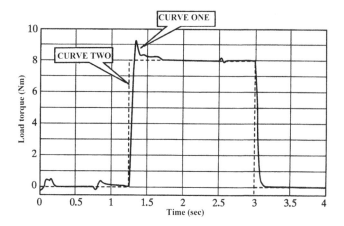

Figure 8.12 Load torque (curve two) and its estimate (curve one) with stator winding inductance increased by 20%

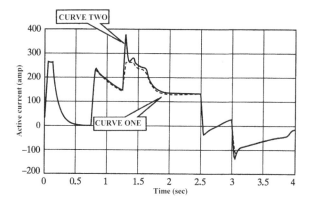

Figure 8.13 Reference (curve one) and actual (curve two) values of the active current by stator winding inductance increased by 20%

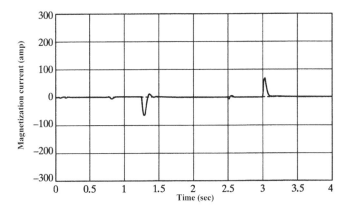

Figure 8.14 Magnetization current at increase of 20% in stator winding inductance. The reference is zero

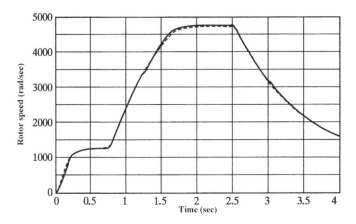

Figure 8.15 Angular speed at 20% reduction of magnetic flux value. The dashed curve is the reference

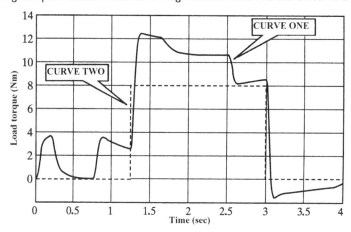

Figure 8.16 Load torque (curve two) and its estimate (curve one) at 20% reduction of magnetic flux value

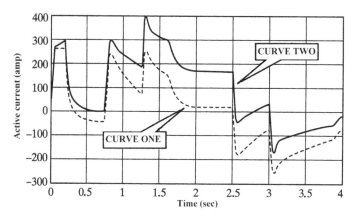

Figure 8.17 Reference (curve one) and actual (curve two) values of the active current at 20% reduction of magnetic flux value

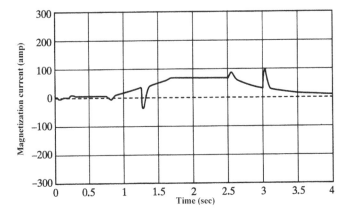

Figure 8.18 Magnetization current at 20% reduction of magnetic flux value. The reference is zero

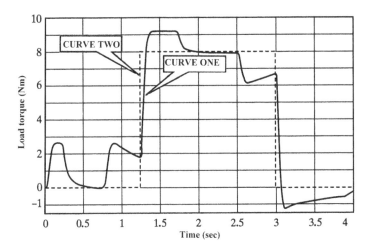

Figure 8.19 Load torque (curve two) and its estimate (curve one) at 20% increase in the moment of inertia of the rotor

oscillograms that a flux change in the synchronous motor (without a corresponding controller updating) also does not lead to loss of stability of the closed loop.

Figure 8.19 shows a oscillogram with the load torque estimation with 20% increase in the rotor moment of inertia ($J = 0.003\,\mathrm{kg}\cdot\mathrm{m}^2$). As we would expect, an error in the reference value of the moment of inertia leads to an error in the load estimation (and to insignificant errors of angular speed control) only in dynamic drive modes, i.e. by changing angular speed.

Parameters rating values of a direct current link are characterized by the presence of a resonance with small enough decrement of attenuation (resonant fluctuations are observed by simulation only if the current controller does not possess sufficient speed). Resonant fluctuations practically don't happen at recommended control rates.

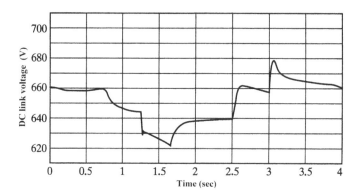

Figure 8.20 Direct current link voltage by tenfold decrease of the inverter input capacity value

From the viewpoint of electrical drive simplification, one important factor is the value of inverter input capacity. A capacity decrease, if it is admissible, allows lowering drive dimensions, simplification of the VSI configuration being near the motor. Thereupon a simulation has been performed at the value of capacity, making 0.1 from rating value (500 µF on one section VSI). Simulation results (direct current link voltage) are shown in figure 8.20. As it appears from the oscillogram, a tenfold decrease in capacity value at the VSI input practically does not affect the process of direct current link voltage variation. That means, obviously, that the energy stored in the capacity is small in comparison with the synchronous motor energy input from the land power source. It means that a considerably smaller capacity of the condenser on an input is required, in comparison with the electrical drive with the asynchronous motor. The choice of capacity value can be carried out starting with the (supply) power capacity, instead of the power (or "control") criteria.

8.1.5   Influence of A/D converter discreteness on current measurements

The problem of insufficient accuracy of an A/D converter of the microprocessor controller is due to its high level of discreteness and can be quite serious. A/D converter discreteness can lead to noise and fluctuations, especially with a high speed controller, or even to loss of stability (self-oscillations). In a sensorless drive the A/D converter discreteness can distort or ruin the work of the computer calculating the angular speed and angular position due to calculation errors. Overcoming the problems caused by A/D converter discreteness, it is possible to use one of following methods:

1) Decrease control rates;
2) Use filters to smooth calculation errors, caused by the A/D converter discreteness. It is important to ensure that such filters do not bring dynamic distortions into the control;
3) Using dynamic models of measured variables change, together with their direct measurements and the subsequent analog to digital transformations.

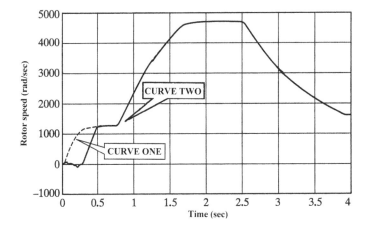

Figure 8.21 Angular speed reference (curve one) and its estimate (curve two) using a 10 bit A/D converter. Curves practically coincide

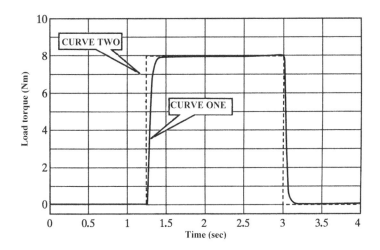

Figure 8.22 Load torque (curve two) and its estimate (curve one) using a 10 bit A/D converter.

The first two approaches are used in the designed control: decrease controller speeds, and the use of filters for the mechanical state variables observer, i.e. angular speeds, angular position and the load torque.

Figures 8.21 to 8.23 present the resulting oscillograms illustrating the influence of the A/D converter discreteness. The system uses a 10 bit A/D converter. A variable of the generalized electric machine is the maximum current amplitude, and it reaches a maximum of 400 amperes approximately. Moreover, a bit is used to take into account the current direction. In other words, the A/D converter will have a discontinuity step value of 0.8 amperes approximately. The resulting oscillograms present the influence of such A/D converter discreteness in the given system, which is considerable, but it is not so essential, even without the corresponding updating of the controller by using a

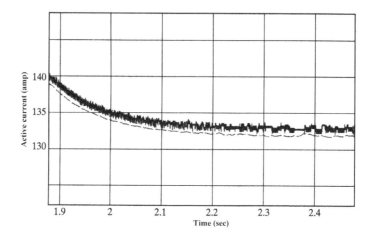

Figure 8.23 Active current using a 10 bit A/D converter. The top curve is an actual motor current after measurement on A/D converter. The bottom one is the current reference (at a larger scale)

change model of the measured currents. It does not lead to stability loss of the closed loop and the process of rotor angular speed control changes only slightly.

The 10 bit A/D converter discreteness becomes apparent at small angular speeds as malfunction of the closed loop work from the resulting oscillograms. That could be related. Thus, it is impossible to provide averaging or filtration of the measured values of currents due to the small feed frequencies of the synchronous motor. The slowly changing phase currents are measured. To expand the control range, it is necessary to use a combination of current measurements (currents observer) with a model of their changes, corrected while measuring them.

8.1.6 The influence of VSI "dead time"

In order to prevent a direct current short circuit by the VSI phase switches switching, the switch-on transistor is made with a certain delay opposite to the switch-off of the another phase switch. The time delay ("dead time") is chosen to ensure that the opposite switch has been closed by any values of current, temperature modes, parameter variations, etc. Depending on a phase current direction by the switch-on delay, one of switching fronts of phase output voltage will be held ("current pause"), while the second one will not. As a result the reference voltage value for the PWM system and the actual VSI realized voltage will differ. It does not represent a special problem in the usual electrical drive due to the presence of an external control loop of a motor mechanical coordinate or other technological variable. In a sensorless electrical drive the information about rotor angular speed and angular position is calculated using values of current and voltage on windings. Usually the voltage sensor is not used. The reference voltage value for the PWM system is taken as information about the winding voltage. In this case, a discrepancy of the reference and the actual voltage value will inevitably lead to errors in rotor angular speed and angular position calculations.

Let us assume that the relative time of realization of each instantaneous voltage vector is not less than the amount of a switch-on delay δ (the latter is identical in all phases). We will assume also that the phase current direction does not change during a switch-on delay. Then the error in the implementation of a voltage vector by any algorithm, the PWM does not depend on a given reference value of the voltage vector and is defined only by current direction. For all algorithms, a PWM current compensation consists in a pause in correcting the reference voltage vector by a constant (for a given direction of a phase current). This vector is proportional to both the ratio of the current pause of the PWM period and the magnitude of the voltage of the direct current link. Its direction is opposite to the signs of the phase currents:

$$
U_{zPWM} = U_z + (U_0\delta/T)\sqrt{2/3}\begin{bmatrix} 1 & -1/2 & -1/2 \\ 0 & \sqrt{3}/2 & \sqrt{3}/2 \end{bmatrix}\begin{bmatrix} \mathrm{sgn}\ i_r \\ \mathrm{sgn}\ i_s \\ \mathrm{sgn}\ i_t \end{bmatrix}
\tag{8.1}
$$

A special case corresponds to the change in the direction of current in phase during "current pauses". During this time both transistors are switched off and the output current flows through the diode. It will be until the moment when the current changes direction. At this moment the diode is switched off. The process can further proceed in the following scenarios:

- After the diode switches off, the voltage value on a phase output is in the range $(0 \cdots U_0)$ of a direct current link voltage and both diodes of the "top" and "bottom" transistor remain in a not conducting state. The output voltage of the given phase is defined by a VSI load, actually, the mode name "pause current" prevails from here, that the current of the given phase remains equal to zero, and the phase potential changes;
- It is required that the phase output voltage has exceeded U_0, or became smaller than zero to keep a zero current in a phase. The corresponding diode of the "top" or "bottom" power switch will open, and actually there will be an expected phase voltage switching.

Something should be said separately about the implementation of the "short" pulse control. If the duration of a control pulse is shorter than the current pause time, then the usual working logic of the control unit PWM, "short" pulses cannot be realized. Error realization of short pulses can be excluded if the operating controller changes the logic of formation current pauses. If the duration of the control pulse becomes less than the pause time, there is no necessity for the time shift on command at the trailing edge of an operating pulse, since the inclusion of an appropriate power switch on the leading edge has not happened yet. Thus, it is necessary to introduce compensation for the control processor "dead" time.

8.1.7 Conclusions on the simulation

The simulation results show that the sensorless electrical drive with synchronous motor remains at nominal (equal) values of plant parameters and the chosen factors of control are stable. It takes place in all the reference range of the rotor angular speeds and the load torque, and in the presence of bounds on variables and controls: a current and a feed voltage of the synchronous motor. Stability degree and control sensitivity are in many respects defined by the established control rates in a controller. By means of a model within the limits of the chosen structure and algorithms, updating controller parameters has been carried out in parts:

-   Filtration degree of the mechanical variables observer;
-   Mechanical movement controller speed;
-   Current controller speed (rates of control).

Drive simulation with the established processes course rates shows the following:

1. 20% motor parameter changes: winding resistance and inductance, magnetic linkage and the inertia moment (without the corresponding updating of a controller) – does not lead to loss of stability of the closed loop, and the rotor angular speed control changes slightly;
2. A/D converter discreteness influence (with 10 bit full range of current phase change) in the given system is considerable, but it is not so critical. The closed loop is stable without the corresponding controller updating by using the model of change of measured currents;
3. Processes simulation by VSI input capacity value making 0.1 from rating value (500.0 µF on one section of the converter) has shown that a tenfold decrease of capacity value practically does not affect the process of voltage change of a direct current link. It gives the chance of a choice of capacity value starting with power capacity, instead of power (or "control") criteria;
4. Discrepancy of the reference and actual voltage at the expense of "dead time" inherent in the inverter leads to occurrence of errors in calculations of rotor angular speed and angular position. It is suggested to enter compensation for the "dead time" in the controller.

8.2 DIGITAL CONTROL SYSTEM OF THE ELECTRIC DRIVE WITH ELASTIC MECHANICAL CONNECTIONS

8.2.1 Control plant features

The given work was carried out together with the state unitary enterprise "Instrument Design Bureau" (State Unitary Enterprise KBP). The goal was a drive control design providing high dynamic and precision characteristics of the electrical drive in the presence of elastic mechanical connections. The given problem is quite complicated due to its high dimension and a considerable presence of a number of nonlinearities, as it was specified in section 7.6. It is more complex than the problem considered in section 7.6, because of two-mass drive construction and the necessity of taking into account the existing backlash in the reducer and the dry friction (Ryvkin et al, 2003).

The block diagram of the mechanical part of electrical drive is presented in figure 8.24. There are two electrical motors having moments of inertia of J_{m1} and J_{m2} respectively. They develop electromagnetic torques on shafts M_{m1}, M_{m2}. The angular speeds of their shafts are Ω_{m1}, Ω_{m2} respectively. The mechanical transfer reduction

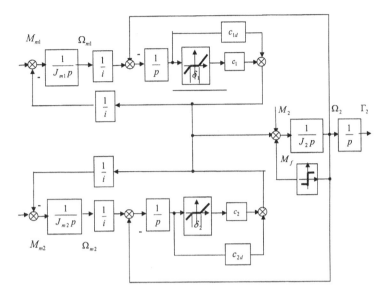

Figure 8.24 Drive mechanical part block diagram

factor from a motor shaft to the load is equal to i. The inertia load (the moment of inertia J_2) connects with the motor through elastic mechanical connections with elasticity factors c_1, c_2 and damping factors c_{1d}, c_{2d} respectively. The transfer backlash values are equal to δ_1, δ_2 respectively. The influence of a backlash and elastic connection is defined by a difference of positions of the load shaft Γ_2, both shafts of motors Γ_{m1}, Γ_{m2} and a difference of angular speeds of the load Ω_2 and motor Ω_{m1}, Ω_{m2}. The load shaft has the load torque M_2 and the dry friction torque M_f. Its sign is defined by the rotation direction of the load shaft.

The design of the microprocessor digital control for the mechanical system with elastic connections is focused on application in the locally distributed digital control.

The processes occurring in such a system are essential nonlinear especially due to the sign changing of load angular speed (backspacing) and by movement into backlash bounds of mechanical transfers.

The mechanical movement system is characterized by the presence of some resonant frequencies that are caused by mass distribution. The lowest (basic) frequency of the mechanical resonance lying in the demanded working strip of frequencies of the closed loop position is usually essential; higher frequencies should be filtered by the control. The control aim is to trace the angular position of the output drive shaft.

The system has the following bounds:

1. A motor angular speed bound due to the bounded motor winding voltage and to strength reasons;
2. A motor electromagnetic torque bound due to the bounded current of the power converter feeding the motor;
3. An output shaft dynamic bound.

Table 8.2 Control design steps (with mechanical movement)

Step Number	Content
1	Drawing up and analyzing the simplified model of the mechanical plant.
2	Working out deference models of the mechanical system for a digital control design.
3	Control design on the basis of simplified deference model. Oscillating motions providing suppression and tracking the reference position of an output shaft. A design method is a modal control of the closed loop with the use of the rate limiter and the observer.
4	Optimization of controller parameters by the analysis of transients, system responses on the sinusoidal input. The closed system pass-band is defined, as opposed to the open system pass-band.
5	Compensation of influence of a dry friction and backslash. Precision characteristics at dumping of load torque. Movement with sign-variable speed of the second mass and movement in a backslash are exactly defined.

The electrical drives also have dynamic bounds. The drive mathematical models, as a rule, have a high order. The control movement and position sensors used are characterized by errors, measurement noise, and, probably, large step-type measurement inaccuracies.

Furthermore, the above mentioned define a highly complex control plant and a control design complexity. Due to the nonlinearities of the plant and the bounds imposed on plant, there are no general analytical methods for control design. It is necessary to accept a number of simplifying assumptions and to use step by step control design. Computer simulations play a big role in checking how the system works in various modes.

8.2.2 Main principles of control design

Control design is done with a cascade control approach. Taking into account that the electromagnetic dynamics of the drive is essentially faster than the mechanical dynamics, it could be possible to simplify the control design problem solution and to decompose an initial control problem into two problems. The first one deals with the motor electromagnetic torque and the second one is the control problem of drive mechanical movement. In this case, for the mechanical system, the motor electromagnetic torque acts as a control. Moreover, basic attention will be given to control design by mechanical movement, since the first problem of the digital control design are considered in detail in chapter 7 and section 8.1. Control design by mechanical movement is carried out in the steps shown in table 8.2.

In order to make it possible to take advantage of the results from section 7.6, the given problem will be reduced to the classical two masses problem, introducing the following variables: $M_1 = (M_{m1} + M_{m2})i$, $J_1 = (J_{m1} + J_{m2})i^2$, $\Omega_1 = (\Omega_{m1} + \Omega_{m2})/2i$, $\Gamma_1 = (\Gamma_{m1} + \Gamma_{m2})/2i$, $k_d = 2c_{1d} = 2c_{2d}$, $k = 2c_1 = 2c_2$.

In this case, the simplified model of a mechanical part (without a backslash. and a dry friction) coincides with the model used in section 7.6. All results from this section for the control design for an electrical drive, with elastic mechanical connections, can be used for the solution of the given problem. The first three design steps are described in section 7.6. It must be emphasized that the use of a reference rate limiter, as described in section 7.5, does not introduce any changes in the control design, since it is an independently designed dynamic system.

8.2.3   Dry friction and backslash compensation

As shown in the block diagram (figure 8.24), the dry friction effect is equal to the load torque. In this case, compensation of the dry friction is reduced using the sum of the load torque and the dry friction in the control. Similarly, it is necessary to change the position value reference of the first mass:

$$\Gamma_{1z} = \Gamma_{2z} + \frac{1}{k_d}\left[J_2\frac{d\Omega_{2z}}{dt} + M_2 + M_f\,sign(\Omega_2)\right] \tag{8.2}$$

For compensation of a backslash it is enough to enter the corresponding correction only into the position value reference of the first mass:

$$\Gamma_{1z} = \Gamma_{2z} + \frac{1}{k_d}\left[J_2\frac{d\Omega_{2z}}{dt} + M_2 + M_f\,sign(\Omega_2)\right]$$
$$+ \frac{\delta}{2}\,sign\left[J_2\frac{d\Omega_{2z}}{dt} + M_2 + M_f\,sign(\Omega_2)\right] \tag{8.3}$$

where $\delta_1 = \delta_2 = \delta$.

The presence of the dynamic processes caused by changes in dry friction force, in general, is inevitable. The dry friction acts directly on the second mass of elastic mechanical system, and it can only be compensated by applying appropriate control of the electromagnetic torque to the first mass, thereby causing the necessary change in the relative position of the masses. For compensation of a dry friction, it is necessary to influence two channels of formation of the motor electromagnetic torque. First, it is necessary to change the torque value. Second, owing to elasticity in the mechanical system, it is necessary to change the position reference of the first mass. The influences of these two channels of friction compensation vary. The additive component in the torque value allows compensating the change of the friction torque at once, as soon as there is a change of a dry friction. However the change of the position reference of the first mass causes error occurrence in its position which, being strengthened by a controller of elastic fluctuations (remember that the controller is characterized by big factors), raises elastic fluctuations in the mechanical system.

Reducing the amplitude of elastic fluctuations in the transition process brought about by changing the sign of the friction torque can be a sufficiently slow change in the job position of the first mass. Thus, it is necessary that the "rate limiter" changes the position reference of the first mass. It must be noticed that similar considerations apply to the dynamic processes caused by changes of the external load torque. The reference rate limiter design is described in section 7.5. However, the value of the compensating additive in the position reference of the first mass is usually insignificant due to the sufficient rigidity of mechanical system. Considering that, and for reasons of simplification of the mechanical control system, it is appropriate for the first order reference rate limiter to specify the position reference of the first mass, by doing it in the form of an inertial link. The position reference of the first mass D_{1z} is formed as follows:

$$D_{1z}(k+1) = D_{1z}(k) + k_D T \left\{ D_{1z}(k) - \frac{1}{k_d}[M_2 + M_f \, sign(\Omega_{2z}(k))] \right\}$$

$$\Gamma_{1z}(k) = \Gamma_{2z}(k) + \frac{1}{k_d} \left[J_2 \frac{d\Omega_{2z}(k)}{dt} \right] + D_{1z}(k+1) \tag{8.4}$$

where k_D is the coefficient defining rates of convergence D_{1z} to the value $[M_2(k) + M_f \, sign(\Omega_2(k))]/k_d$.

There is a qualitative change of mass behavior in the backslash movement. The backslash leads to the termination of mass interaction. The first mass moves only under the influence of the electromagnetic torque and the movement inertia, while the movement of the second mass is influenced by the external load torque and a friction and the mechanical system "is disconnected". With a disconnected system, generally speaking, the design problem statement of elastic fluctuations a controller loses its meaning. It is necessary to solve other control problems. If, with backslash movement, a speed of relative movement of masses increases, after passing this zone, there is an impact of masses, there are shock loadings on connection elements that are inadmissible for reasons of mechanical durability and decline of a working resource. Thus, the possibility of staging the control problem when moving the mechanical system in the backslash consists in exiting from that zone with a small (close to zero), finite rate of relative movement at the restoration contact of masses.

The information on mutual position of the first and second masses is necessary for the solution of the problem. Such approach is basically possible, but even under the smallest backslash, the possibility of estimating the mutual position of masses inside this zone gets complicated. It is due to the limited precision characteristics of the position sensors and also to the reducer errors. Finally, because of discrepancies in the knowledge of the structure and plant parameters, it is desirable to find a simplified solution, which would allow to bind the speed of relative movement of masses in these conditions and to lower the shock torque to a comprehensible level. One of the simplified solutions lies on the surface and follows from above. For the analysis of the dynamic processes, it is necessary to limit the intensity of change of the position reference of the first mass, caused both by passing band backlash, as well as by changes of both moments, dry friction and load.

Taking into account all that, it is necessary to add the rate limiter of the position reference of the first mass:

$$D_{1z}(k+1) = D_{1z}(k) + k_D T \left\{ D_{1z}(k) - \frac{1}{k_d}[M_2 + M_f \, sign(\Omega_2)] \right.$$

$$\left. -\frac{\delta}{2} sign \left[J_2 \frac{d\Omega_{2z}(k)}{dt} + M_2 + M_f \, sign(\Omega_{2z}(k)) \right] \right\}$$

$$\Gamma_{1z}(k) = \Gamma_{2z}(k) + \frac{1}{k_d} \left[J_2 \frac{d\Omega_{2z}(k)}{dt} \right] + D_{1z}(k+1)$$

$$\tag{8.5}$$

The demanded value of factor k_D is specified at the simulation.

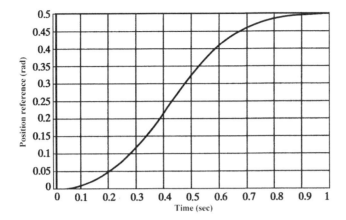

Figure 8.25 Position reference for mechanical movement controller before and after the reference rate limiter. Big amplitude step response

Thus, the system, in which the static and dynamic errors arise from load torque action, dry friction and backlash, are compensated by design. Sufficiently simple and logical simplifying assumptions imply that the values of the operating electromagnetic control torques in the transition process are limited. Such a system seems to be rather unrefined to inaccuracies in measuring the used variables.

8.2.4   Closed loop simulation

The simulated system consisted of the model of the mechanical system with elastic connections with two drives and a massive output mass, a reference rate limiter, a controller of mechanical movements, a load torque reference system and a graphic display of simulation results. The presence of a set of resonant frequencies (a high order plant) was not considered at the simulation. The resonant frequency of the mechanical fluctuations in the opened loop is about 10 Hz, the elasticity factors k is $4 \times 10^7 \text{N} \cdot \text{m} \cdot \text{sec/rad}$, the damping coefficient k_d is $16 \times 10^6 \text{N} \cdot \text{m/rad}$, the backlash value δ is one milliradian, the moment inertia load J_2 is $8000 \text{ kg} \cdot \text{m}^2$, the reduction factor is 170. The electromagnetic drive torque was limited at level of three rating values. The angular speed was limited at level 1.2 from nominal. The use of two motors allows applying a choice mode backlash in "tightness" (with different electromagnetic torque references).

Software packages MatLab (version 5.2.1) and Simulink (version 2.2) were used for simulation. Mainframes were simulated using the C++ language. Blocks of mechanical system miscalculate with a "small" step, allowing investigating the processes in continuous time. Control system blocks (controller, rate limiter) are considered discrete, with a period of 1 millisecond.

Figures 8.25 to 8.27 show the rate limiter response (position, speed, acceleration) for a step system input of big amplitude (0.5 radians). The speed and acceleration do not reach restrictions in transient under a small step input, as opposed to a system response to a big one.

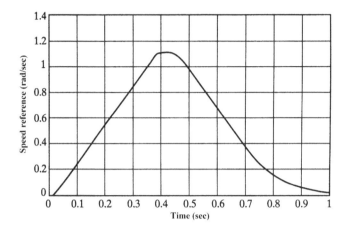

Figure 8.26 Angular speed reference for mechanical movement controller after a reference rate limiter with big amplitude step response

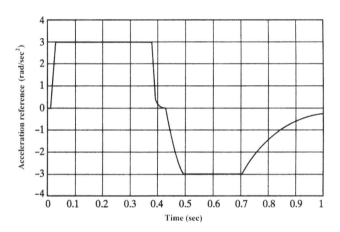

Figure 8.27 Acceleration reference for a controller of a mechanical movement controller after a reference rate limiter

Figure 8.28 presents the rate limiter response to the additive system input: step of a 10 milliradians and the small amplitude sinusoidal signal of 10 Hz frequency. The rate limiter in the established mode practically does not make dynamic distortions on such input frequencies.

Figure 8.29 shows the control plant output error by step position reference with amplitude of 50 milliradians.

Figure 8.30 shows the control plant output by sinusoidal reference of amplitude of 0.2 radians and a frequency of 2.5 Hz.

Figure 8.31 shows a control plant output error in less detail (one division of a vertical scale corresponds 0.05 milliradians). The error hits in the servo mode could be explained by backslash and dry friction influence.

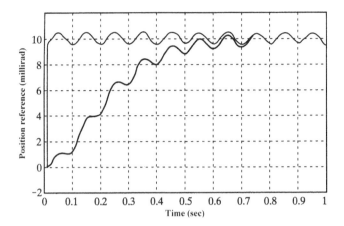

Figure 8.28 Reference rate limiter output with an additive input. Step one is 10 milliradians and a 10 Hz small amplitude sinusoidal signal

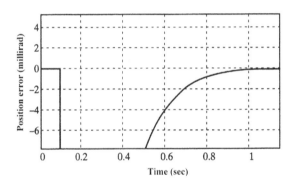

Figure 8.29 Control plant output error with a step position reference

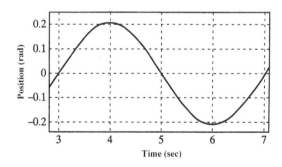

Figure 8.30 Control plant output by the sinusoidal position reference with frequency of 2.5 Hz and amplitude of 0.2 radians

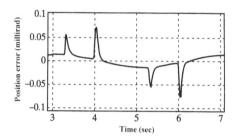

Figure 8.31 A tracking error by the sinusoidal position reference with frequency of 2.5 Hz and amplitude 0.2 radians

It appears from the simulation results, that the suggested digital control of an electrical drive with elastic mechanical connections provide high quality of control processes, even in the presence of dry friction in the system and backslash in mechanical transfers.

REFERENCES

Ryvkin S., Izosimov D. and Baida S. "*Flex mechanics digital control design*". Proc. IEEE International Conference on Industrial Technology, IEEE ICIT'03, 2003, pp. 298–303.

Ryvkin S., Izosimov D., Sarychev A., Raskin L., Aksarin D., Vidumkin E. and Cernat M. "*Sensorless drowned oil pump drive*". Proc. IEEE International Symposium on Industrial Electronics, ISIE, 2005, pp. 963–968.

Index